Tim Spector is Professor of Genetic Epidemiology at King's College London and Hon Consultant Physician at Guy's and St Thomas' Hospital. He has won several academic awards and published over 700 academic papers, a large proportion of which relate directly to nutrition and the causes of obesity. Since 2011 he has been leading the largest microbiome project in the UK, using genetic sequencing to study the bacteria in the guts of 5,000 twins, and in 2014 he became the lead investigator of the British Gut Project at King's.

www.tim-spector.co.uk
@timspector

Also by Tim Spector

Identically Different
Your Genes Unzipped

Tim Spector

THE DIET MYTH

The Real Science
Behind What We Eat

WEIDENFELD & NICOLSON

A W&N PAPERBACK

First published in Great Britain in 2015
by Weidenfeld & Nicolson

This paperback edition published 2016
by Weidenfeld & Nicolson,
An imprint of the Orion Publishing Group Ltd,
Carmelite House, 50 Victoria Embankment,
London EC4Y 0DZ

An Hachette UK company

1 3 5 7 9 10 8 6 4 2

A CIP catalogue record for this book
is available from the British Library.

ISBN 978 1 780 22900 3

Typeset by Input Data Services Ltd, Bridgwater, Somerset

Printed and bound by CPI Group (UK) Ltd, Croydon, CR0 4YY

www.orionbooks.co.uk

To my family and other microbes

Contents

A Bad Taste

It had been a tough climb: six hours walking up 1,200 metres to the summit on touring skis with artificial sealskins to stop us sliding backwards on the snow.

Like my five companions I was feeling tired and a bit light-headed but I still wanted to check out the spectacular view at 3,100 metres over Bormio on the Italian–Austrian border. We had been ski-touring in the area for the past six days, staying in high-altitude mountain lodges, enjoying plenty of exercise and good Italian food. We took our skis off to walk the ten metres to the top but I felt unsteady and didn't go all the way to look over the edge, thinking my mild vertigo was kicking in. As we turned to ski down, the weather deteriorated, clouds descended and light snow began to fall. I had trouble seeing the tracks ahead of me but assumed it was my old goggles misting up. Usually skiing down is the easy relaxing part, but I was strangely tired and relieved an hour later when we reached the bottom.

When I caught up with our French mountain guide, he pointed out a large tree fifty metres away with two alpine squirrels in it. I could see the squirrels, but I could see four of them – two diagonally above the others – and realised I was seeing double. From my days as a junior doctor in neurology I knew the three likely causes at my age, none of them good: multiple sclerosis, brain tumour or stroke.

After a stressful few days back in London when I managed to or-ganise an MRI brain scan, which, luckily, didn't show anything that suggested the two other unpleasant causes, I was still left with the possibility that I'd suffered a small stroke.

Eventually, an ophthalmologist colleague was able to diagnose me over the phone with a fourth cranial nerve occlusion. I had only vaguely heard of it, but the good news was that it usually improved within a few months without treatment. The exact cause is unknown but it involves a spasm and constriction and micro-blockage of the artery supplying this nerve, which in turn controls some of the eye movements. It was a great relief. I just had to wait for the eye to return to normal and wear initially a patch and then some nerdy-looking glasses with prism lenses to help reduce the blurring.

I couldn't read or use my computer for more than a few minutes at a stretch and, to complicate things, I had developed high blood pressure. This puzzled my expert colleagues as blood pressure is not supposed to change so suddenly, but I knew mine definitely had, as, by chance, I had measured it myself two weeks before. After many cardiac tests to exclude rare causes I was given anti-hypertensive drugs and aspirin to thin my blood.

In the space of two weeks I had gone from a sporty, fitter-than-average middle-aged man to what felt like a pill-popping, hypertensive, depressed stroke victim. With the enforced time off work as my vision slowly improved I had plenty of opportunities for contemplation.

This was the wake-up call I needed to reassess my own health, and sent me on a personal odyssey not only to understand how to improve my chances of living longer and better but also to reduce my dependence on prescribed drugs and find out if by altering the food I chose to eat I could become healthier. I thought changing my lifelong dietary habits would be my greatest challenge – but it turned out that finding out the truth about food was an ever greater one.

The myth of modern 'diets'

Trying to work out what is good or bad for us in our own diets is increasingly difficult, even for me as a doctor and scientist who has studied epidemiology and genetics. I have written hundreds of scientific papers on different aspects of nutrition and biology but have found it hard to make the shift from general advice to practical decisions. Confusing and conflicting messages are everywhere. Knowing who and what to believe is a big problem. While some diet gurus tell us to 'graze' by eating regular small meals and snacks, others disagree and encourage, say, skipping breakfast, eating a big lunch or avoiding heavy meals at night. Some promote eating one thing (such as cabbage soup) to the exclusion of others, while there's a French diet cleverly called *le forking* which claims that by using only a fork to eat the pounds will fly off.

Over the past thirty years almost every component of our diet has been picked on as the villain by some expert or other. Despite this scrutiny, globally our diets continue to deteriorate.[1] Since the 1980s, when the links between high cholesterol and heart disease were first uncovered, the idea that a healthy diet has to be low-fat has taken hold. Most countries have reduced their official recommendations for the amount of total calories consumed as fat, particularly meat and dairy products. Reducing fat meant increasing carbs. This has been the mainstay of medical advice and, superficially at least, seemed to make sense, since fat packs twice the amount of calories per gram as carbohydrates.

In contrast to this official line, diet plans of various levels of complexity such as the Atkins, Palaeolithic and Dukan Diets, which have become popular since the early 2000s, all urge people to stop indulging in carbohydrates and to eat only fat and protein. The glycaemic index (GI) diet targets certain types of carbohydrates that via the release of glucose rapidly raise blood insulin, seen as the main enemy, and the South Beach Diet targets both bad carbs and bad fats; some diets (such as the Montignac) forbid certain food combinations, and the recent phenomenon of fasting (such as the 5:2 diet) promotes as the answer intermittent 'fasting' via periods of reduced calorie intake. And there are countless alternatives – I was shocked to find well over thirty thousand books available, with their own websites and merchandising, promoting different diet regimes and supplements ranging from the sensible to the dangerous and crazy.

I wanted to find a formula that would keep me healthy and reduce the risks or symptoms of the most prevalent modern diseases. But most popular diet plans focus on reducing weight rather than other health and nutritional aspects. Some people are overweight yet suffer few adverse metabolic consequences, while others are apparently lean with little fat under their skin but have fat around their inner organs, with disastrous consequences for their health. But scientists still don't understand why this happens.

The ritual of dieting has become an epidemic. A fifth of the UK population are on some form of diet at any one time, yet we continue to expand our waistlines by an inch every decade. The average male and female Briton now has a 38-inch and 34-inch waist respectively and both are still increasing, leading to more and more related health issues like diabetes and knee arthritis and even breast cancer, the rate of which increases by a third for each increase in trouser and skirt size. While 60 per cent of Americans would like to lose weight, only a third actually bother any more – a significant reduction from twenty years ago. The reason is that most people don't believe the weight-reducing diets actually work. Surrounded by increasingly plentiful and cheap food, and harsh memories of failed attempts at dieting, we often lack the willpower to reduce our calorie intake and to exercise more. There is even some evidence that an endless cycle of failed diets, where weight drops and rebounds regularly, can actually make people fatter. Some of the popular diets clearly work for many of us in the short term, especially the low-carb, high-protein ones, but longer term it seems to be a different story. The evidence suggests that even with record-breaking dieters, the weight often slowly piles back on.

Bad science and increasing waistlines

Since the 1980s the experts have been consistently telling people that eating any amount of fatty foods is bad for us. This campaign has been very effective, and with the help of the food industry has managed to reduce the total amount of fat consumed in many countries. Despite this, rates of obesity and diabetes have increased even faster. We have since discovered that some of the most prolific consumers of fat in the world, the Cretans from southern Greece, are among the healthiest and longest-lived. In order to replace the fat content, the food industry has steadily increased sugar levels in processed foods. This has led to dire warnings of sugar being the arsenic of our times. Yet it turns out to be still more complicated. Cubans, despite eating on average twice the total amount of sugar as Americans, are poorer but far healthier.

It is not then surprising that we are confused by all these different and competing messages – avoid fizzy drinks, sugar, juices, fat, meat, carbs – and left feeling as though there is nothing left to eat except lettuce. This confusion plus counter-intuitive food subsidies on corn (also called maize), soya, meat and sugar explains why people in Britain and America are actually eating less fruit and vegetables than a decade ago, despite expensive and aggressive government campaigns. In Britain, 'five a day' advice was recently stepped up to 'seven a day', in a futile attempt to stem the tide in the opposite direction. The reasoning behind these and most official diet recommendations is obscure – the simplicity of the message is overriding science. And there is little cross-national consistency. Some countries make no recommendations, others have now moved to 'ten a day'; and others like Australia proclaim 'Go for two and five' so as to distinguish fruit from vegetables and to stop people just drinking seven orange juices a day. The food industry loves these ideas, and adds 'healthy' labels to their processed foods to obscure the other elements.

The justification for the 'seven a day' advice in the UK was based on an observational study of 65,000 people which compared those who said they had eaten no fruit and veg at all the day before to those who had eaten over seven portions. The survey reported that eating fruit and vegetables reduced death rates by over a third but that the absolute death rate could be lowered by just three-thousandths (or 0.3 per cent) in the fruit and vegetable eater – not so impressive. Genetic factors or, more likely, social factors could explain the food preferences, especially given that someone in East Glasgow is likely to die twenty years earlier than someone living in affluent Kensington. A study ten times larger found no benefit in increasing portions past five a day.

I'm not saying the advice is always wrong, but when it comes to health and diet we need to be much more cautious and critical of 'official' advice and recommendations. These knee-jerk responses are often based on insufficient evidence or bad science, or simply a reluctance on the part of politicians and scientists to change tack for fear of 'confusing' the public and losing face.

Just as dangerous is the oversimplification of the 'common sense' approach. If you eat less and exercise more you will lose weight, and if you can't manage this you simply lack willpower, goes the message. This has been another medical mantra for the last few decades. Despite increased longevity, more sophisticated medical technology and improved living standards, we are going through an unprecedented epidemic of obesity and chronic ill health with no obvious end in sight. Can this really be due to a global lack of willpower, as we are often led to believe?

Many of the British twins I study have been put on diets, and it has been interesting to see how they have fared compared to their twin who hasn't tried the same diet. When we asked them whether they had ever been on a weight-reducing diet for over three months, the ones that replied yes were on average fatter than the ones who said no. So to try to achieve a fair comparison of the effect of dieting rather than of the different personalities or physical characteristics of the twins, we looked at the differences between twin pairs. This let us account for any difference in genes, upbringing, culture and social class, which for most twins were perfectly matched. We also selected for this study only identical twins where both individuals in the pair were overweight, with a body mass index of over 30 (BMI is calculated as your weight in kilograms divided by your height squared in metres). For medical and research purposes, doctors classify this as obese.

At the start of this experiment the average weight of these twelve highly selected female twins was 86 kg (13½ stone) and their average BMI was 34. Now, you might have predicted that the twin who had the willpower to diet regularly would have something to show for her years of sacrifice. Instead, I found absolutely no difference in weight between the twin who had dieted regularly for the past twenty years and her identical twin who had never been on a serious diet. Similar results were found in younger twins who started off at the same weight at the age of sixteen. The twin who had dieted was on average 1.5 kilos heavier when the two were compared at the age of twenty-five.[2]

Our bodies simply seem to adapt to the new reduced calorie intake and do what they are programmed by evolution to do. It appears that the dull monotony of most exclusion diets is overridden by the body's impulse to hold on to our fat stores. Once someone has been obese for a while, a whole series of biological changes transpire to maintain or increase their fat storage and the brain's reward mechanisms for food.[3] This is why most diets fail.

Global time bomb

In 2014, over twenty million American kids were obese – a percentage of the population that has tripled in three decades. Even American babies, who clearly can't be blamed for their willpower or lack of it or for making poor choices, are getting fatter at a frightening rate. And the rest of the world is catching up: in the UK two out of three adults are now overweight or obese; the Mexicans are now unofficial world champions, and have overtaken the US in rates of both childhood and adult obesity; in China and India rates have tripled in thirty years to almost one billion obese citizens; over one in ten children in countries whose populations are often assumed to be thinner, like Japan, Korea and France, are now classified as obese.

Obesity, although sometimes seen legally as a disability, is not classified as a disease, yet its effects are just as deadly. As well as costing countries billions in healthcare bills, the main health consequences of this epidemic now include diabetes, which affects over three hundred million people and is growing at the rate of 2 per cent per year – double the average population growth rate. In places such as Malaysia and the Gulf States almost half the population have diabetes. If current trends continue, by 2030 an extra seventy-six million people in the UK and US will be clinically obese, bringing the obesity totals close to half the population. This means millions of extra patients with heart disease, diabetes, stroke and arthritis. Taxpayers are the ones footing the astronomical bills incurred, while we are being told by our governments and doctors that they know exactly what the problem is: overeating.

But why does the number of obese people on the planet continue to sky-rocket in developing countries like Botswana and South Africa where nearly half of all women are now clinically obese, while thirty years ago we were predicting mass starvation due to lack of food?

My earliest personal encounter with the extreme consequences of obesity was in the 1980s while working as a junior doctor in the first ever obesity unit in Belgium. To begin with, my junior colleagues and I jokingly regarded it as an expensive health farm. My first patient changed all that. She was brought in by the fire brigade, having collapsed at home with a blood clot in the lung. Weighing 260 kg

(40 stone), she had been too heavy for the ambulance, and had to be winched out of her window by the fire crew. At only thirty-five years old, a diet of junk food and soft drinks had led to her being trapped in her own home for years, gaining weight until her body broke down. Despite losing 100 kilos in hospital she continued to suffer a series of severe medical problems including diabetes, arthritis and heart disease, and she died two years later of heart and kidney failure.

At the time of this first encounter with obesity, in 1984, the condition was still extremely rare. When I saw the effects on that real person and patient my view of obesity and its consequences altered completely. Such sad stories are now quite common, like the Welsh teenager from Aberdare who weighed 56 stone and had to be rescued from her home by demolishing a wall.

When I returned to Britain it would be another twenty years before doctors took the rise in obesity at all seriously, and even today obese patients are routinely denied treatments, compassion and resources. They cannot get urgent operations, and all over the world they are treated as second-class citizens when it comes to healthcare. Obesity is still a massively neglected area of medicine, with little funding, no speciality training, and no common voice with which to try to combat the billion-pound marketing budgets of the food companies.

As a junior doctor in London I was regularly told by my consultant bosses to tell obese patients with major health problems to exercise, to 'take control of their lives and use their willpower to stop overeating', or perhaps to remind them that 'there were no fat people in concentration camps'. Needless to say these not-so-subtle 'medical' methods failed miserably – my patients got progressively fatter, more depressed and more diabetic and disabled. Sometimes we referred them to the hospital dieticians, but this was always a futile exercise as they were simply asked to change their habits and stop eating biscuits and crisps. It was like trying to use a sticking plaster to treat a massive haemorrhage. What was needed was a total change of approach.

If you reduce the daily calorie intakes of overweight people for long durations in a controlled environment to fewer than 1,000 calories (our normal recommended intakes are 2,000–2,600 calories a day) you have the solution to obesity. However, outside the army or hospitals such conditions are impossible, and there remain no practical or proven effective cures. One artificial exception, which also 'cures' diabetes without changing the external environment, is radical gastric bypass surgery. Yet despite fifty years of its relatively safe use doctors are highly unwilling to recommend it, partly because they don't understand why it is so effective.

Doctors, dogma and diets – reversing ignorance

When faced with my own health scare up in the mountains, my knee-jerk reaction was that I must give something up. I chose to give up meat and dairy and the saturated fats that go with it, but depending on whatever article I had read last it could just as easily have been carbs, grains, e-additives, gluten, pulses or fructose. As the twentieth-century story of how all fats are bad for us seemed to be unravelling, I wanted to uncover the real science behind this and other diet myths. I wanted to find out if there was something all the so-called experts were missing.

Was I right to give up meat, which humans have eaten for millions of years? Do milk, cheese and yoghurt really cause allergies as many studies now claim? Was I eating too many carbohydrates or grains to compensate for the lack of fat and protein? Should I worry about GI content of carbs? The truth is that generally in science or medicine the yes-or-no answers favoured by doctors and other health experts turn out to be wrong. There is nearly always another layer of biological complexity and control that either hasn't been thought of or has been dismissed as unimportant. This book is about digging down to that next layer using the very latest scientific research.

As well as my own experiences to draw on, I was lucky enough to have a large research group of fifty people and 11,000 adult twins I had been studying for over twenty years to help me. Being able to separate the effects of diet and environment from the effects of our genes is one of the big challenges of the nutrition research world, and twins offer the solution. These volunteers from all over the UK have been providing us with information on their health, lifestyles and diet habits in extraordinary detail. Combined with all the genetic data we have on them and their co-twins, they are probably the most studied people on the planet. This book has been an extraordinary personal tale of discovery for me, and I hope it will help you to cut through the confusing dogma, commercial interests and diet myths that face us all.

I want to use the latest research and discoveries to reverse the trend of ignorance, and to think outside of what is currently a very tightly closed box. I want to demolish the myth that obesity is simply a matter of counting calories in and out or about eating less and exercising more or cutting out one food type. It can seem today as though everyone is an expert on food and diet. But most diets are designed or promoted by people with no scientific training, and sadly, although there are some sensible ones, anyone can call themselves a nutritionist or nutritional consultant. Famously, a professional certification from the American Association of Nutritional Consultants was awarded to one Henrietta Goldacre. The fact that Henrietta happened to be the medical author Dr Ben Goldacre's deceased cat demonstrates the high standards of many nutritional diplomas.[4]

Even respected doctors become entrenched in their ideas and theories and refuse to acknowledge their flaws when new data emerges to contradict them. No other field of science or medicine sees such professional infighting, lack of consensus and lack of rigorous studies to back up the health claims of the myriad dietary recommendations. Moreover, no other field of science feels to me so much like a mass of competing religions – all with their high priests, zealots, believers and infidels. And as with religion most people, even at the risk of death, are unwilling to change their faith.

With the nutrition professionals constantly contradicting and criticising each other, it is no wonder that few large collaborative studies or projects get funded. I know from personal experience that many academics seeking funds for a project deliberately omit to mention an important diet component because they know that it will be heavily criticised by colleagues. Although there are a huge number of small studies performed and paid for each year, the standard of research compared to other fields is lagging far behind. Most studies are still cross-sectional and observational, full of possible biases and flaws; a few are superior observational studies followed over time, and only a tiny fraction are the gold-standard randomised trials in which subjects are randomly allocated to one foodstuff or diet and followed for long durations.

What we continue to lack is a wider understanding of the science behind nutrition and diets. Most diets are based on a narrow traditional view or simple observation and quackery, but the massive differences between individuals and their physiological responses to food go unexplained. If each new processed food introduced into our diet were a drug made by a pharmaceutical company, and if obesity were labelled a disease, we would have a wealth of data on its benefits and risks. Yet with food, even for the most synthetic chemical concoctions, we have no such safeguards.

The missing piece of the jigsaw

There is an enormous piece of the nutrition puzzle missing. Why can one person eat a certain meal regularly and gain weight while another ingests exactly the same food and yet loses pounds? Lean people (by which we now mean those of healthy weight and a BMI less than 25) are now the minority group in most populations. What makes them so different from 'normal' overweight people? Perhaps we should be studying them as the 'abnormal' ones?

Some of these differences are clearly down to our genes, which influence both our appetites and our eventual weight. My studies of twins in the UK (the TwinsUK study) and others from around the world have shown that identical twins are much more similar to each other in body weight and fat than fraternal twins. Since they are effectively genetic clones and share the same DNA, this shows the importance of genetic factors, which explain around 60–70 per cent of the differences between people. On average, identical adult twins are less than 1 kilo different in weight. These gene-influenced similarities extend to other, related, characteristics that we have also explored, such as the percentages of total body muscle and fat, and exactly where fat is deposited in or on the body. Habits related to eating are also influenced by genes, such as food likes and dislikes, and even how often people like to exercise or have meals. However, just because a trait is 60 or 70 per cent 'genetic' doesn't mean it is predestined.

In fact, identical twins can sometimes have very different waistlines despite having identical genes, and we are studying these special pairs in great detail to find out why. These genetic factors alone don't explain the massive changes seen in the population over the last two generations. In the UK in 1980 only 7 per cent of men and women were obese – now it is 24 per cent. Genes, made up of variations in DNA, cannot change that fast and traditionally need a minimum of around one hundred generations to adapt by natural selection.

Clearly, other factors are involved. The twenty-first century has already achieved major breakthroughs in the genetics of obesity and brain chemistry, and these newly discovered genes certainly do play a role, but it is only a very minor one. It is possible that we have been ignoring another major factor that influences our diet and health: this is our tiny gut microbes that may hold the answer to our modern obesity epidemic.

I'll introduce microbes in detail in the next chapter, as they are crucial to so much of our misunderstanding about our modern diet. This fascinating new area of scientific investigation is transforming our entire understanding of the relationship between our bodies and the food we consume. Our narrow, blinkered view of nutrition and weight as a simple energy-in and energy-out phenomenon and our failure to account for our microbes have been the main reasons for the miserable failure of diets and nutritional advice. This nutritional disaster, combined with our success in mass-producing ever-cheaper food and in treating some diseases, is allowing us to survive longer, but at the cost of making us increasingly unhealthy.

Armed with this new science, we need to rethink our approach to food, nutrition, diet and obesity. In the twentieth century we came to see food in terms of its component (macronutrient) parts that provide the energy source – protein, fat, carbohydrate and so on. We have all become used to seeing these listed on food labels, and much medical and nutritional advice is based on this vast oversimplification of food in all its complexity. I want to show why this approach is a mistake. I don't want you to stop taking medicines or following diets prescribed by your doctor, but I do want you – and your doctor – to question their underlying rationale. Taking the nutritional information on typical food labels as a guide, I want to show why we now need to look beyond the superficial advice we find on them. Along the way I hope to expose and demystify many of the most dangerous myths about modern diets.

1

Not on the Label: Microbes

If I told you about a fellow creature that shares our food and habits, travels with us, has evolved with us to know what we like and dislike, and that we provide protection for, you might assume I was talking about your beloved dog or cat. In fact I'm talking about something a million times smaller and invisible to the naked eye.

Microbes are primitive forms of life that were the first inhabitants of earth, creatures that we generally ignored or took for granted. These creatures, too small to be seen with our eyes, we assumed to be mainly found in dirt and in or on other animals that didn't wash. Yet our bodies contain 100 trillion of them, weighing over four pounds in our guts alone. Most of us know them intimately only from their associations with rare bouts of food poisoning – like salmonella in uncooked barbecue chicken or E. coli in an unwise late-night kebab. Apart from occasions like these, with all our ever-expanding knowledge and technology we assumed that these tiny and seemingly trivial beings couldn't possibly influence our supremely powerful human bodies. We couldn't have been more wrong.

Dancing animalcules

Spring, 1676: Anton Leeuwenhoek had overslept again and it was already light when he woke. There was noise and activity in the streets of Delft below. He had worked long into the night on his latest experiment and was still tired, but elated by his recent discoveries. Using his special home-made microscope, Anton had been looking at why chilli peppers are spicy, but by chance had stumbled across something even more revolutionary.

Anton was a draper by trade and was obsessively curious. Unlike most of his friends, he still had his own teeth and was fastidious about cleaning them daily, first by rubbing them vigorously with hard salt granules, then using a wooden toothpick, then rinsing, before polishing them with his special tooth cloth.

Today he took particular interest in examining with his fine magnifying mirror the whitish batter-like substance (nowadays known as plaque) that was coating his teeth. Anton had only a small amount of plaque compared to others he had examined, but even after he had cleaned his teeth it never seemed to have gone away completely. He scraped some off onto a glass slide, then added a few drops of fresh rainwater. Upon examining the slide Anton was amazed. There were tiny wriggly creatures everywhere. These 'animalcules', as he named them, were of all shapes and sizes – at least four distinct families of them, all 'a-dancing prettily'. What shocked him was not their presence, but their abundance. 'The number of these Animalcules in the scurf of a man's teeth are so numerous that I believe they exceed the number of men in a kingdom,' he wrote.

Anton Leeuwenhoek was perhaps the first man ever to see a microbe (by which we mean a living creature seen only by means of a microscope). He was certainly the first to describe them, and to realise that healthy humans are teeming with these creatures inside our guts and on our skin. He found them everywhere he looked, from our mouths to our food, from our drinking water to urine and stool samples. Despite this amazing discovery, unlike Newton and Galileo – scientists of the same era who explored outwards to the stars to achieve fame – he slid into relative anonymity.

You may not have given microbes much thought until now, perhaps because you can't see them without the aid of a magnifying glass. Imagine how many grains of sand there are on earth – or, if you prefer, how many stars there are in the universe. Someone has actually counted the stars – well, made a very good estimate – and came up with a figure of 10^{24} (which is 1 + 24 zeros – an awful lot). If you multiply the estimate for all possible stars a million-fold you get a vast figure which, at 10^{30} (also called a 'nonillion'), is the estimated number of bacteria on earth. If you are a gardener and by accident swallow a tiny fleck of earth it contains billions of bacterial cells, and a handful of earth contains more microbes than stars in the galaxy. You are no 'safer' swimming in the water, where a million bacterial cells are found in every millilitre of fresh or sea water. These microbes are the true and permanent inhabitants on earth; we humans are just passing through.

Microbes are present in most habitats – from the ordinary to the most extreme. Bacteria inhabit acidic hot springs, radioactive waste, and the deepest portions of the earth's crust. Bacteria have even survived in space. We evolved not from Adam and Eve but from microbes, and we have continued our close connection with them ever since. This is most obvious in our guts, where thousands of diverse species that are as different from each other as we are from jellyfish play a much greater role than we ever realised.

Microbes generally get a bad press, but less than a tiny fraction of the millions of species are harmful to us and most, in fact, are crucial to our health. Microbes are not only essential to how we digest food, they control the calories we absorb and provide vital enzymes and vitamins as well as keeping our immune system healthy. Over millions of years we have evolved together with microbes for mutual survival, yet recently this fine-tuning and selection has gone wrong. Compared to our recent ancestors who lived outside cities, enjoying rich and varied diets and without antibiotics, we have only a fraction of the diversity of microbial species living in our guts. Scientists are only now starting to understand the long-lasting impact this has had on all of us.

Early colonists on virgin soil

Our personal encounter with microbes begins at birth. Within minutes of a healthy sterile baby's delivery she will be swarming with microbes: millions of bacteria and even more viruses that feed off the bacteria, plus even a few fungi. Within hours she will be totally overrun with millions more.

Her head, eyes, mouth and ears are the first parts to be colonised as she passes along her mother's soft vaginal wall, where many eager microbes in the moist and warm mucosal layer are waiting to make the leap. Then, because of their close proximity and the pressure on all of the body's sphincters, a light mixture of urinary and faecal microbes are sprinkled onto her face and hands, followed by a different set of microbes covering the rest of her body as a result of rubbing against the skin of her mother's legs. These tiny microbes are carried to the lips and mouth, usually from the baby's own hands. They can't usually get past the oceans of saliva sweeping them away, and if they do they face the harsh acid environment of the stomach and its juices, where most are destroyed.

At the first swallow of some alkaline breast milk (which acts like an antacid), a few lucky bacteria waiting either on the lips or mouth or on the mother's nipple will be miraculously protected, and make it past the acid waterfall. These intrepid explorers can then start a whole new colony by reproducing wildly in the safety of the mucous layers of the baby's intestine, and wait for more milk and other microbial companions to arrive. Even just a few colonists – if the conditions are right – by dividing every 40–60 minutes, can become trillions or billions of cells overnight.

Until the mid-1990s it was the dogma that most bodily fluids were sterile – that is, contained no microbes. When a team in Madrid claimed to have cultured dozens of microbes from healthy breast milk they were laughed at.[1] Now we know that human milk contains hundreds of species, although we still don't have a clue how they get there. We are no longer sure that any part of our bodies is completely devoid of microbes – even the womb and the eyeball – and they may even travel around our bodies unnoticed.[2] When you next go to the toilet, spare a thought for your trillions of microbes. Nearly half the mass you are flushing away are microbes.

Although we are all born microbe-free, this state lasts just a few milliseconds. The process of microbe colonisation is not at all random and has been planned and finely tuned over millions of years. In fact both the microbes and the baby depend on each other for their survival and health. This delicate co-evolution between microbes and man has not left purely to chance the crucial planting of the first microbial seeds in the virgin soil. All mammals, and many other animals studied such as frogs, transmit their own carefully selected microbes to their babies in a process that is at least fifty million years old. This is how evolution has facilitated the leap of microbes from one generation to the next, and how our own unique community of microbes, called our microbiome, is established.

Diverse microbial gardens

We are surrounded by trillions of microbes in the dirt, dust, water and air that are not interested in colonising a newborn baby. They haven't yet evolved the apparatus to be able to survive on or inside us and to derive enough energy to live. So the microbes that colonise humans are highly specialised, possessing pared-down genes to ensure there are no redundant or overlapping mechanisms with the human host. We humans share 38 per cent of our genes with the microbes inside us. As the transmission of microbes from mother to child is so universal in the animal kingdom, it is clearly crucial for our health.[3]

As soon as a woman gets pregnant the body readies itself for providing as much help as it can for the next generation via this special transmission of microbial genes. Inside the pregnant female the carefully programmed changes brought about by switching on genes in the body ensure that certain hormones modify metabolism and calorie intake while conserving energy, building up fat reserves in the breasts and buttocks, increasing glucose and stocking up breast milk. Other changes happen to the white cells controlling the immune system that must deal with the foreign object – the baby – inside her, without rejecting it. There are also changes to the microbes, anticipating the day when they will be transmitted to her baby in order to aid his or her growth and survival. These microbe changes are extremely powerful.

When researchers transplant the pregnant human mother's stools into sterile mice, the mice get much fatter compared to those transplanted with non-pregnant human faeces.[4] Experiments using these sterile, or germ-free, mice are a vital tool that we scientists use regularly in this area of research. They are carefully brought into the world via a sterile C-section in an oxygenated enclosure, avoiding contact with the other littermates or their mother or other microbes. They are then kept in sterile and isolated cages, fed sterile food, and observed. Without microbes they can survive, but only just. They definitely do not form the elite corps – they are puny, and don't develop a normal brain, gut or immune system. Most significantly, they are expensive to feed as mice without microbes need a third more calories than normal mice to keep up their body weight – evidence of the vital importance of microbes in digesting food in the intestines.[5]

Most of our microbes inhabit our large intestine (the colon), which is the five-foot-long piece of gut before you get to the rectum and that absorbs most of the water. The piece of gut above it – the small intestine – is where most of our food and energy is absorbed into our blood system. Usually, food enters here having first been chopped up by our teeth, aided by the enzymes in our saliva and stomach. The small intestine also contains microbes, in smaller numbers, but we know much less about them and their precise roles. High-fibre food, which needs more time to break down to release the nutrients, gets sent on from here down to the large intestine filled with microbes.

If you give the sterile mice normal microbes after a few weeks they still never develop normally, but if they start life with gut microbes and you try to eradicate them with antibiotics (as humans sadly do all too often, with disastrous effects), although never healthy, they do much better.

Microbes predict obesity better than genes

Recent changes in our tiny gut microbes and to the community called our microbiome are likely to be responsible for much of our obesity epidemic, as well as its deadly consequences of diabetes, cancer and heart disease. Examining the DNA of the microbes in our guts gives us a much better predictor of how fat someone is compared to looking at all of our 20,000 genes. This predictive ability is likely to keep improving as we start to look at viruses and fungi as well. Subtle differences between the types of microbes that we host in our guts explain many of the links between our diets and health, and why the results of food research are so inconsistent between people and populations. For example, differences in our individual gut microbes can explain why a low-fat diet works for some people, while a high-fat diet is fine for some and dangerous for others; why some people can eat plenty of carbohydrates without problems and others extract more calories from the same amount and get fatter; why some eat red meat happily and others contract heart disease; or even why when old people move to care homes and their diets change, they often rapidly succumb to diseases.

The increasing promotion and use of restrictive diets that depend on just a few ingredients will inevitably lead to a further reduction in microbe diversity and eventually to ill health. Intermittent fasting (such as the Fast Diet or 5:2) may be the exception, as short-term fasting can stimulate friendly microbes, but this is only as long as the other, 'free eating', days contain a diverse diet. Fifteen thousand years ago our ancestors regularly ingested around 150 ingredients in a week. Most people nowadays consume fewer than twenty separate food items and many, if not most, of these are artificially refined. Most processed food products come depressingly from just four main ingredients: corn, soy, wheat or meat.

In 2012 I started the world's then largest gut microbe study (Microbo-Twin), using the latest gene technology and five thousand twins to identify the microbes and their relation with diet and health. I subsequently launched the British Gut Project, which is a crowd-funded experiment linked to the American Gut Project that allows anyone with access to the internet and a postal service to test their own microbes and share the results with the world.[6] I also experimented on myself with some diets and will share with you the exciting insights these bring to a new vision of nutrition. Only by understanding what makes our own personal microbes tick and interact with our bodies can we make sense of the total confusion of modern diets and nutrition and regain the correct balance of our ancestors.

A 2015 study of microbes in all the stations of the New York subway found they matched closely their previous human hosts – the diverse population groups of the city, each with their own idiosyncrasies. The study also found that half the microbes it discovered were completely unknown.[7] The good news is that although there is clearly still much to learn, we already know enough scientifically about our microbes and our bodies to enable us to alter our lifestyles, eating patterns and diets to suit our individual needs and improve our health.

It is useful to think of your microbial community as your own garden that you are responsible for. We need to make sure the soil (your intestines) that the plants (your microbes) grow in is healthy, containing plenty of nutrients; and to stop weeds or poisonous plants (toxic or disease microbes) taking over we need to cultivate the widest variety of different plants and seeds possible. I will give you a clue how we do this. Diversity is the key.

2

Energy and Calories

Eat less and exercise more and you will burn off calories and lose weight. I, like most doctors, used to say this regularly to my patients. Experts tell us that the reason we have increased our weight so dramatically in recent years is that we have become more sedentary and are consuming more food. Put another way, people become fat because they eat more calories than they expend. On the face of it, this reasoning seems hard to contest.

Fixating on the basic laws of thermodynamics – energy in must equal energy out – has distracted us from the questions of how and why. We don't say that someone becomes alcoholic simply because they drink more alcohol than they can metabolise. We are obviously interested in the reasons some people but not others become alcoholic in the first place. Yet we are happy to say that obese people are fat simply because they eat more calories than they use up, without asking why.

The misleading medical calorie dogma

A calorie is a calorie – this is the central tautological dogma of traditional diet and nutritional advice. At a basic level this statement is correct. A calorie is defined as the amount of energy liberated when a standard unit of dried food is burnt off. The phrase means that regardless of the food type the calorie comes from (protein, fat or carbs), the energy needed to extract it and the energy produced will be identical. This has been the basis of calorie-counting for decades. It forms the basis of our food labels which many people use to make nutritional choices. But what if this lab-based approach has been misleading us into thinking we understand nutrition and diets?

One real-life study exposed some of these fallacies. Forty-two monkeys were fed two different diets with identical calories for six years in controlled conditions. The ingredients were identical except for the fat content: one group had 17 per cent of total calories based on natural vegetable oils; the other had 17 per cent based on artificial and unhealthy trans fats. The diets were designed to keep weight constant, but the trans-fat group gained weight and accumulated three times the harmful visceral (belly) fat and much worse insulin profiles (meaning that glucose in the blood is not disposed of quickly) compared to the other group.[1] This suggests that all calories are not the same. Two thousand fast-food calories will have very different energy consequences from 2,000 calories made up of whole grains, fruits and vegetables.

For too long we have taken the accuracy of our food labels for granted, but the formulas behind them are over a hundred years old. They depend on burning the foods and applying calculations to account for different rates of digestion and absorption. The formulas disregard the effects of how old food is and also the different effects of cooking, which can dictate how much is absorbed as well as the speed glucose can rise in the blood. Also, people with longer large intestines can extract more calories from food than those with short ones, and some studies have shown differences of up to 50 centimetres between populations.

These formulas are just based on estimated 'averages' in a nonaverage world. Errors have been found that overestimate calories for foods such as almonds by over 30 per cent, and manufacturers are legally allowed error rates of up to 20 per cent on their product labels.[2] Many common items, such as processed frozen foods, underestimate calories by up to 70 per cent, and high-fibre products by 30 per cent; and while health claims are scrutinised by regulators, in most countries there is astonishingly little oversight of the accuracy of nutritional labels.

To add to these errors, there is even more uncertainty around the average numbers of calories men and women need each day to replace energy lost. Recent recalculations have increased the gold-standard averages to 2,100 per day for women and 2,600 for men. Many people believe that this is too much – for a start, the guideline figures obviously fail to account for age, height, weight or amounts of activity.

Clearly, the use of calorie counting in diets depends not only on the accuracy of the system, but also on people's ability to estimate correctly the number of calories in their food. Studies consistently show that only about 1 in 7 people comes close to estimating the number they need. The idea that the source of calories is unimportant can also lead to major imbalances in protein, carbohydrate and fat intakes, noticeably high or low levels of which can have serious health consequences. In America, restaurants and cinemas are now required to provide calorie counts on their menus. The evidence that they will help customers is unclear, although it may force manufacturers to reduce the number of calories in new products.[3]

How the body produces energy from food varies enormously depending on the source of the food, how much you chew it, how easy it is to digest and what else you eat with it. One study even showed that eating white rice with chopsticks rather than a spoon significantly reduced the speed at which the blood glucose rose and triggered insulin (called its glycaemic index (GI)).[4] Many experts believe that this GI score for food is crucial to regulate weight, although the few controlled clinical studies performed in humans have so far failed to show any difference in weight or heart risk factors when high- and low-GI diets are compared directly.[5] But response to calories also depends on your own physical and genetic make-up and, last but not least, on the microbes in your gut. A clever, detailed study of 800 Israelis in 2015 explored their blood sugar responses to identical foods. They found fourfold variations which were much more related to their own gut microbes than the type of carbohydrates or GI index.[6] None of these factors are considered when food is reduced to calories or misleading and often irrelevant GI scores on a food label. So a calorie may indeed be a calorie, but in the real world inside our intestines they are definitely not equal in the effects they have.

The fatted calf and the 3,600-calorie diet

Jerome was one of twenty-four student volunteers in Quebec who took part in a unique research study in 1988. It was a dream summer job: nearly unlimited free food and accommodation for three months – he would also be paid, and all in the name of science. He had passed the qualifying test, showing he had no family history of obesity or diabetes, and he was of normal height and weight. He, like the other volunteers, was a typical healthy but slightly lazy student who did no regular sport. Once he had signed his consent forms and waivers he found himself prisoner in a specially rented campus dormitory sealed off from the outside world, where for the next 120 days he was to eat, sleep, play video games, read and watch TV. There was twenty-four-hour supervision, and no sport, alcohol or smoking was allowed during the study except for a thirty-minute walk outdoors every day.

Jerome's first two weeks involved daily weigh-ins, food question-naires and immersion in a water tank to calculate his body fat. He was similar to the other subjects who were all on the skinny side, weighing only 60 kg (132 lb), with a normal healthy body mass index of 20. He was taken to a dining room where for each meal there was a buffet and a choice of foods. Every bit of food on the plate that he ate was carefully weighed. His baseline intake was calculated over two weeks and found to be 2,600 calories on average. After the run-in period he and the other volunteers were then overfed with an additional 1,000 calories a day for 100 days, under strict conditions so that they couldn't cheat and pass food to each other. They were given a regular diet of 50 per cent carbs, 35 per cent fats and 15 per cent protein. Jerome was measured and scanned at the beginning and end of the study.

After the hundred days on this 3,600-calorie diet and virtually no activity, his weight had increased by 5.5 kg (13 lb). When the researchers looked at all the students' results they were surprised to see such a wide range of weight gain. Jerome was the second-lowest weight gainer – some of his colleagues had gained an impressive 13 kg (29 lb) in the same time. The only student that had gained vir-tually the same weight as him over the three months was Vincent, who happened to have been born in the same town, went to the same school and shared all his genes. In fact, he was his identical twin brother. Dr Claude Bouchard, a professor at Laval University, Quebec, and his colleagues had cleverly selected twelve pairs of vol-unteer twins. They all gained weight to very different extents, but in each case their weight gain was very similar to their twin's.[7] Al-though all the twins gained total weight and fat mass, some details varied too. Some pairs converted the calories not just into fat, but into additional muscle. They also seemed to gain the fat in the same places as their twin, around the belly or more unhealthily around the intestines and liver – what is called visceral fat.

This classic study, in which the students were overfed like lab rats, might now have trouble getting ethical approval (though we don't protect actors like Bradley Cooper who gained 40 lb for the film *American Sniper* and was paid millions of dollars for his role in it). The twins study unequivocally shows that much of how quickly we use energy or store fat and so gain weight is clearly down to our genes. My studies of thousands of twins in the UK and other studies around the world have consistently shown that identical twins – who, as mentioned earlier, are genetic clones – are much more similar to each other in body weight and fat than are fraternal twins, who share only half the same genes. This again shows the importance of genetic factors, which can explain around 70 per cent of the differences between people. Furthermore, we found these similarities extend to other, related, characteristics such as how much muscle or fat you have, and where the fat is stored on the body.[8] No one knows what sends signals to our fat cells to expand typically around our stomachs and buttocks, and why we don't, for instance, have chubby elbows.

Individual habits to do with eating (for instance, whether you are a grazer or a gorger) are not just picked up by watching your family or friends eat well or badly: they also have genetic components. This includes a like or dislike of certain foods, such as salads, savoury snacks, spices and garlic. How often you take regular exercise was also shown by our twin consortium to have a strong genetic component right across the world.[9] A combination of novel experimental and cross-national twins studies have shown that people with genes for obesity also have genes that make them less likely to exercise than naturally skinny people, thereby highlighting some of the extra pressures obese people are under when they try to slim. Their genes and bodies conspire against them when they try to burn off calories.

Thrifty genes

For a long time, the best theory to explain why we were rapidly getting fatter was the 'thrifty gene' hypothesis of the 1960s.[10] The idea was that in the last thirty thousand years or so (that is, in our recent past, since our ancestors left Africa) we survived a number of major events that dramatically reduced populations through illness or starvation, such as mini ice ages or enforced long trips to find food. An example is the Pacific Islanders, who sailed thousands of miles across the ocean to find both food and more hospitable lands. En route many would have perished. The theory goes that those who could best build up their reserves beforehand and then retain their fat on the voyage were more likely to survive (sometimes by eating the skinny ones). That fat protects against starvation is well documented.[11] So when the diminished population eventually arrived on their paradise islands, the skinny ones had been weeded out and subsequent generations were highly selected for fat-retaining genes.

This appeared to make sense, because some of the most obese humans on the planet come from the islands of Nauru, Tonga and Samoa, who became fat only recently when their environment changed and they were exposed to an abundance of easy food and little incentive to exercise. The high death rates of transported African slaves en route to the US offer us another example often used to explain the higher risk of obesity in African-Americans today. Any differences in obesity between countries could then be explained by their stage of development from food scarcity to food abundance. So the theory proposes, in fact, that we are all descended from just a few families who supposedly survived famines or climate changes. Many of us have therefore inherited variations of genes that at some point in the past were a great advantage, but now are definitely not.

There are major flaws in this theory, however. First, it assumes that for most of their lives our ancestors had only just enough food to survive and would rapidly gain weight if faced with a surfeit. But the idea that they were always running out of food and rarely had any excess is probably incorrect. Studies of current and past hunter-gatherers suggest that most of our ancestors generally took in plenty of calories. This makes sense, as humans have also lived in travelling groups of fifty to two hundred who varied widely in size, age and food needs. This meant that if most of the time they had sufficient food to feed the biggest and most needy, the rest must have had food in excess.

The thrifty theory also assumes that protection from starvation was the main explanation for these genes being selected by evolution. But it is more likely that deaths from childhood infections and diarrhoea, as in the developing world today, rather than famine, were the major evolutionary drivers. Increasing body fat whether in children or in adults is not a strong protection against infections.

The other myth is that all our ancestors ran around all the time looking for food, like hyped-up ultra-marathon runners. Although it's likely that some were keen runners, the same studies of hunter-gatherers suggest they would have been resting or sleeping much of the day and didn't consume many more calories overall than we do now. Other studies have shown that wild animals brought into captivity and faced with abundant food don't suddenly become obese. Finally, every population group studied has skinny exceptions to the rule. Even where the 'normal' state now is to be obese and diabetic (like the Pacific Islanders or the Gulf States' populations), there is always at least a third of the population that manages to stay lean, despite being surrounded by an abundance of cheap calorific food or lazy compatriots. Such increasingly rare slim individuals may become the best groups to study.

Drifty genes

These holes in the 'thrifty gene' theory led a British biologist, John Speakman, to propose a rival model of obesity, not as well known, called the 'drifty gene' hypothesis.[12] The idea is that until two million years ago our genes and our mechanisms for retaining body fat were more tightly controlled, and that we would have had a major survival problem if we had been too chubby. Old skeletons of our ancestor the Australopithecus show many signs of having been regularly eaten by hungry predators. Some of these species, such as Dinofelis, a variety of not-so-cuddly 120-kilo sabre-tooth tiger, even specialised in hunting early humans. Being fat meant not only that you couldn't run as fast and so would be easier prey, but that you would taste better than a stringy marathon runner. These are two very good reasons why those obesity genes in our distant past would be selected against, and our upper fat limit controlled.

Becoming too thin, however, was of course always a disadvantage.

Although food was generally plentiful, everyone needed fat reserves for emergencies in the days before fridges and freezers. So at both ends of the thin–fat spectrum our genes masterminded mechanisms to push us back towards the middle road. As we evolved into *Homo sapiens* with our bigger brains, hunting and weapons skills, we lost our fear of predators but still had occasional threats of famine and changes in climate to contend with. This maintained our tight genetic controls for minimal levels of fat, particularly in the useful storage areas. Many women, in particular, know personally how tough it is to shift those last bits of fat from their bottoms or thighs, despite dieting and months in the gym.

Gradually, though, as our natural predators disappeared so did our need to run away quickly. Consequently, over the last million years or so the genetic controls over the upper threshold of body fat have become more relaxed. While some people might by chance have kept these genes, in others the gene effect has weakened and the threshold has drifted upwards. This means that some of us will keep increasing our fat levels up to this raised variable threshold, and others – around a third of the population – will stay lean even if surrounded by food.[13] This also makes sense as the genes for leanness overlap with those for increased physical activity.[14]

Another popular misconception is that in recent decades thin people have become fat. Studies of obesity trends have confirmed that in the last thirty years of global obesity epidemics most lean people have stayed lean; it is the slightly chubby that become obese and the obese that become very obese. There does seem to be a higher threshold or upper limit for most people: most, once they reach a certain weight, and regardless of the extra quantities they then eat, cannot get much fatter.

Surveys of twenty-five countries carried out between 1999 and 2009 show that some (but not all) Western countries may be finally starting to hit this upper fat threshold: there appears to be a flattening of the obesity curves, particularly in children and adolescents.[15] In the USA, where the epidemic started, the adult figures for obesity have for the first time started to flatten (but not drop).[16] However, for obvious reasons this is not widely advertised – having only a third of the population staying clinically obese is hardly a success story. Paradoxically, Americans may be relatively genetically protected compared to Asian populations. Judging by the speed with which they are catching up, and their tendency to lay down visceral belly fat, Asian populations may have even higher upper thresholds, and could keep growing outwards for even longer.

Good taste and supertasters

The ability to taste has been called our nutritional gatekeeper. People who completely lose their sense of taste don't get fat. We all have up to ten thousand taste buds on our tongues for five main tastes: sweet, bitter, sour, salty, umami (a savoury taste related to monosodium glutamate (MSG)). We may even have a sixth one called kokumi, meaning heartiness. Contrary to the myth, our taste buds are not separated and we can taste the different tastes across the tongue. The buds regenerate every ten days and are controlled by genes influencing their relative sensitivity. The differences in our genes produce the variations in how sensitive we are to certain foods and how much we like bitter or sweet tastes.

Taste genes probably evolved so that as we travelled and encountered an increasing number of plants we would be better at detecting edible ones containing nutrients, while avoiding those that were toxic. The considerable variations in sensitivity may perhaps have evolved so that whole tribes would not be wiped out by eating the same poisonous fruit. In 1931 a Dupont chemist found out by chance in his lab that 30 per cent of people cannot taste a substance called PROP that 50 per cent of people find bitter and 20 per cent intensely unpleasant. Here was clear proof that our taste experiences are unique.

We probably have hundreds of different taste genes, and more variants are being discovered every year. Most of those found so far belong to two gene families (called TAS1R and TAS2R). There are at least three gene variations for sweet-taste detection (for fruit), over five for umami (as a marker of protein), and at least forty for bitter tastes (toxins). Which gene variants we have influences not only our appreciation or dislike of foods but also our fat, vegetable and sugar intakes. The bitter and sweet receptors are also present in our noses and throats and unexpectedly play a role in signalling to our immune systems when to expect a microbial infection. These taste receptors malfunction when you get abnormal continuous infections, such as sinusitis, that overload the system.[17]

When it comes to bitter tastes, a small proportion of the population are so-called supertasters. This means they have odd variants of one of the TAS2R genes and they react strongly to the chemical PROP in tiny dilutions. These individuals are very sensitive to strong flavours and tend to be much pickier with their food. The taste genes make supertasters sensitive to subtle differences in many nutritious vegetables, like those of the brassica group that includes cabbage and broccoli, as well as green tea, garlic, chillies and soya. As a result they tend to avoid some of these vegetables, often dislike drinking beer and other alcohol, and find cigarettes too bitter. With their discerning palates, although they miss out on some good foods they are usually healthier and less likely to be fat.[18]

As food types vary in their calorific content, food preferences in omnivores with plenty of choice can play an important part in determining their energy and weight. In 2007 we performed a twins study combining UK and Finnish twins to explore why some people prefer sugary foods to others. We found that nearly 50 per cent of the differences between people who had a sweet tooth and those who didn't were due to their genes, and the rest was down to culture and environment.[19]

Gene variants for greater sweetness sensitivity (TAS1R) are much more common in Europeans than in Africans or Asians, suggesting that Northern Europeans evolved these genes to help them detect new food sources as they moved away from the safety of the equator. The ability to tell by taste whether a new root vegetable was edible and nutritious had clear survival advantages when faced with inconveniences such as an ice age. Unfortunately these same genes do not help us to survive in the aisles of a modern supermarket. Most studies suggest there is only a weak association between having these sweet-tooth genes and increased body fat.[20] It used to be thought that you were either a sweet or a savoury person. In children at least, this idea has been dispelled by a recent study showing that the liking for salt and the liking for sweetness go together – and as kids prefer both sugar and salt at higher levels than do adults, they are particularly vulnerable to being exposed early to the modern processed food diet.[21]

Exercise and willpower

Are we really doing less exercise? We have talked about calories that are simply units of energy produced when foods are burnt as fuel, and calories that are eaten but not burnt off as body fuel are stored as fat. But what is the role of exercise in expending calories? If you are trying to get fit and healthy, exercise works – you don't need a fancy meta-analysis to prove that. Even the experts and the nutritionists can agree that regular exercise improves your heart and muscles and increases your lifespan. They don't yet agree on how much exercise you need, but the range is somewhere between 90 minutes and six hours per week of moderate activity, enough to work up a sweat. Others disagree with that, suggesting just a few minutes a day of flat-out running or cycling in the form of a short sharp shock is enough to fool your body that it's getting a good workout.[22] The beneficial role of gentle walking is even less clear, but it's probably still better than doing nothing.

Exercise is not, however, merely a question of willpower. A few years ago we combined the major twin cohorts of Europe and Australia and looked at the exercise habits of nearly 40,000 twin adults. After the age of twenty-one when the influence of parents and family starts to wear off, a liking for participation in leisure exercise several times a week, in every country, was around 70 per cent heritable – that is, highly genetic.[23] This shows that exercise is much easier for some people than for others; their bodies and minds find the process more pleasurable than do those others who may feel nauseous even when just watching sports on TV. Clearly, people and their bodies can change, but the starting positions may be very different.

As with recalling meals and diets, smoking and drinking, we have unreliable memories where our exercise habits are concerned, and we tend to exaggerate. One way round this is to use activity monitors, new instruments that correlate your heart rate with movement detected by sensors. These monitors calculate daily activity very accurately and reveal how many of us overestimate it. They also show the enormous variation between individuals and how some people move about and fidget even at rest, which also expends energy. Some studies have suggested that a tendency to fidget is a useful protection against obesity. Some fidgeting genes have been found in mice which are also active in the human brain, resulting in some restless people expending up to 300 calories per day more than a restful person.

We tested our twins with an acti-heart activity device. They wore this trendy type of wristwatch for a week as it recorded their pulse and their activity. The results proved what we already knew – that there is a clear 70 per cent genetic component in self-reported sports. But, more surprisingly, the genetic component of actual energy expenditure was below 50 per cent for most measures, and around 30 per cent for the act of 'sitting around'. This means that environment is slightly more important than genes in your real energy expenditure.[24]

Some studies, rather than focusing on exercise, have looked at sitting on your bottom as a risk factor. Regardless of how much exercise you do (or claim to do), the hours you spend watching TV or sitting in a car are independently a risk factor for heart disease and mortality. Large observational studies in the UK and the US have shown that for every two hours of TV viewing per day your risk of heart disease and diabetes increases by 20 per cent, even after accounting for other risk factors.

My father didn't watch much TV but he spent his life avoiding exercise. He was brought up at a time when many people thought that exercise was bad for you. He was naturally very skinny and my grandmother made huge efforts to build him up. He would jokingly say to us as kids, 'I used to be a nine-stone weakling, now in middle age I'm a twelve-stone weakling!' He loathed parents' school sports days and usually found an excuse not to participate. He couldn't run as he had flat feet, couldn't skate, ski or ride a bike as he had no balance, and couldn't swim as he had heavy bones. He claimed to be descended from a long line of Jewish non-sportsmen and women.

We tend to forget just how recent is the latest craze for fitness and sports. In the 1980s joggers in strange pyjama-like outfits were seen as weirdos and treated with derision. The New York marathon started with just over 100 participants in 1970 and the London equivalent began ten years later, but by 2015 nearly one million people have crossed the finish line. In the early twenty-first century the number of adults who do some kind of gym or sporting activity is large and growing. In 2014, 13 per cent of adults in the UK were members of a gym or exercise facility and many more trained outdoors in parks or took part in team sports. And more than a third of British over-fifties do some regular gardening.

The UK gym business is worth nearly £3 billion a year and the US has over fifty-one million gym members, the business having grown nearly twentyfold since the 1970s; and there is a similar picture in most countries. But if we are actually exercising more, shouldn't we be getting thinner, not fatter? – unless most of us just go to the gym to watch TV, sit about in the jacuzzi and drink smoothies – a good way to get fatter without guilt?

Can it be true, as we are often told, that despite all this leisure activity we really are much more sedentary than we were thirty or forty years ago? Our jobs may have become less manual thanks to labour-saving devices, but our leisure time is more likely to incorporate exercise. However, if work-related exercise used to be important in preventing obesity, why are manual workers, who expend more calories in their jobs, consistently more obese than office workers? Part of the problem is that accurate calorie-expenditure data is hard to collect and compare over the decades, leaving us with very few hard facts to rely on.

One long-term study of housewives living in Minnesota has shown that life has got easier for many of them. They observed major shifts in the amount of energy they expended daily on household chores as compared to sedentary behaviour such as watching TV. Compared to 1965, fifty years later they now apparently expend 200 calories fewer per day.[25] However, more detailed and representative survey data collected from the Netherlands between 1981 and 2004 shows that while over time body fat has increased significantly, leisure exercise levels, which might have been expected to have diminished, have actually slightly increased.[26] Another review of several studies in the US and Europe since the 1980s found that, in contrast to popular perception, there was no overall difference in total daily energy expenditure including working time, and physical activity has not declined.[27]

Exercise and other physical activity are consistently linked with the strength of bones and muscles, which in turn has been associated with changes in the rates of osteoporotic fractures – especially hip fractures, which affect one in three women. In the 1980s a couple of colleagues and I examined the changing rates of hip fracture in the US and the UK over forty years for which we had accurate data. What we saw was that, adjusting for age and demographic changes, US fracture rates increased dramatically until the mid-1960s, then tailed off. In the UK they also increased after 1950, then plateaued in the 1980s; and according to my colleagues doing further analysis they have not increased further since.[28] The results were a surprise to us then, but they now fit with the evidence that, contrary to the received wisdom, our overall level of exercise hasn't changed much since the 1970s in the US or since the 1980s in the UK.

Does exercise really help you lose weight?

The standard advice from dieticians and gym instructors is that if you burn off around an extra 3,500 calories through exercising you will burn off a pound of fat. The 'go for the burn' motto certainly helps motivate gym junkies. But the energy expenditure of most people's weekly sweaty gym-class workout equates only, sadly, to the reward of a large doughnut afterwards.

To compensate for the hours on end I have spent sitting unhealthily on my bottom writing this book, I have also been trying to train for a triathlon. I thought this would mean expending some serious calories. While on sabbatical in Barcelona I enjoyed the luxury of being able to swim around a mile in the sea every day and cycle forty to sixty miles in the surrounding hills at weekends. I walked for about thirty minutes a day and ran occasionally (in between some annoying injuries). I estimated with the help of my GPS sports watch that I was expending an extra 3,500 calories per week on average, and I wasn't aware of eating more than usual. In ten weeks I lost barely 1 kg (about 2 lb), far from the impressive 10 lb I should have lost if the mythical fat-calorie formula were correct – which it clearly is not.[29]

My experience, though anecdotal and unreliable, is not unique. In one study, 12,000 regular runners who subscribed to the US *Runner's*

World magazine were followed over many years and the number of miles they ran per week was tracked to their weight each year. Although there was found to be a correlation between distance running and leanness, nearly everyone – however far they ran – still slowly got fatter each year. The authors suggested that if you added an extra four to six kilometres a week to your run every year you might, if lucky, stay the same weight, but you'd eventually need to be running sixty-plus miles a week.[30]

The reason why millions of us don't lose weight exercising is that our bodies compensate. The body is programmed to stop us losing weight via fat and we have to expend five times more energy to get rid of fat than muscle.[31] It may convert some of the fat to muscle – but that doesn't show up on the scales. As children we were told to go outside and play so as to work up an appetite, and this was also for another reason. It made us hungrier the next day too and slowed the body and its metabolism down in subtle ways. Careful exercise studies, in which sedentary volunteers exercised intensively for six months, found they lost only 1.5 kg in weight as opposed to the 4.5 kg expected. Their hunger and food intake did increase, but only by 100 calories a day, which wasn't enough to explain the failure to lose weight.[32] Many other exercise studies show that energy expenditure when at rest stays low or that it drops by up to 30 per cent with more exercise. This reduction is mainly due to a drop in metabolic rate or in subconscious movements like fidgeting, which also expends calories.

If exercise alone does not lead to significant weight loss, when people have successfully lost weight in three to six months through diet, can exercise work to keep it off? The short answer is no. In a recent meta-analysis of seven studies exploring exercise alone or exercise plus diet versus diet alone, exercise failed dramatically to have any effect over placebo or control interventions. Nearly everyone regained weight, and without dietary restriction exercise had little influence.[33] [34]

Fit or fat?

So is it worth exercising if it doesn't help you reduce weight? There is an interesting debate going on about whether it is better to be thin and sedentary or fat and fit. The studies are pretty consistent: being fat yet fit is definitely better than being thin and unfit for heart disease and for overall mortality. The major risk factors for heart disease associated with being unfit – smoking and not eating vegetables – outweigh the risk of excess body fat. A study following up over 300,000 Europeans found that doing no exercise whatsoever carried twice the risk of early death as obesity. Just doing twenty minutes per week brisk walking for a totally sedentary person (which is over one in five Europeans) would reduce their risk of premature death by a quarter.[35] So it is very important to get the right overall balance for health even if you are overweight. The exception to the rule is the risk of diabetes, where being thinner consistently reduces your risk, even if you are unfit and do no exercise.[36][37]

My father was not fat and didn't smoke but he was very unfit and had a fatal heart attack at the age of fifty-seven; so there is a lesson there, even though some people (like my father) will find it tougher than others to overcome their anti-sports genes. Exercise is overall a pretty good investment in time for most people – around 270 hours of annual exercise adds around three years to your lifespan and delays the onset of many diseases.

Our microbes are born to run

Our microbes certainly play a part in how exercise can reduce our risk of disease and early death, but the mechanism is currently poorly understood. Exercise stimulates the immune system in beneficial ways, then the immune system in turn sends chemical signals to the microbes in our guts.[38] But it could also work the other way round, as exercise alone can also influence the gut microbiota composition directly.

One experiment was done on gym rats (real ones). Healthy rats love to run, and when divided into those with a running-wheel in their cage and those without, the runners, who averaged 3.5 km per day, produced twice the rate of the beneficial short-chain fatty acid butyrate in their guts compared with the sedentary rats.

Butyrate is a small fatty substance produced by our gut microbes that has many beneficial effects on the immune system, and exercise stimulates microbes to produce more of it.[39] Having the right kind of gut microbes may also make you run faster or swim further, possibly owing to the microbes' antioxidant properties. Antioxidants are important chemicals that prevent the release of substances called free radicals from cells – substances that cause a series of chain reactions shortening the life of the cell. So antioxidants are considered healthy chemicals, and are contained in many foods and produced by microbes. Perhaps altering your microbes will become the latest doping trend in the Olympics – although only elite long-distance swimming mice have been caught cheating so far.[40]

In the American Gut Project and our twin study, which are both cross-sectional observational studies, the strongest factor found to date affecting the richness of the gut microbes in over three thousand people was the amount of exercise they reported performing. However, in this kind of study it is hard to separate this from other associated factors, such as healthy eating. The best human data so far comes from a unique study showing the growing interest in microbes in the elite-sports nutrition world. Many elite sportsmen and women are now having their microbes profiled and their diets modified by their nutritionists.

In one study the stools of elite athletes of the national Irish rugby squad were sampled during their intensive pre-season training.[41] Forty of these beefy men had a mean weight of 101 kg and a BMI of 29 – showing, incidentally, that about 40 per cent of them were technically obese and the rest overweight (but you probably wouldn't want to tell them that yourself). In reality, you would be hard pushed to find any body fat on them (they had average levels of 16 per cent, which is very low). This emphasises how unreliable BMI is, and its weakness in measuring obesity in populations where waist–hip ratios or even belt size may be more effective measures. The researchers tried to find a comparable group, but of course it proved impossible. They found twenty-three men of the same age and BMI from Cork, but their extra BMI came mainly from fat (33 per cent), not muscle. So as an extra comparison they found another group of skinny local men.

The results showed clear differences: gut microbiota diversity was significantly higher in the athletes compared with both the other groups. The rugby players who consumed more calories also had healthier inflammatory and metabolic markers and greater numbers of most microbes. Microbiota-diversity measures positively correlated with the higher protein intake and markers for extreme exercise. By picking such an extreme elite group, the study couldn't really separate the effects of exercise from those of diet, but suggested that both diet and exercise were driving the changes in microbial diversity. The bottom line, however, is that although exercise is not of much benefit for your weight or for burning fat (unless you are a professional athlete), it is good for you, your heart and your longevity. And since it also makes your microbes healthier and more diverse, it is a good thing.

Brain food

For those of you who genetically or culturally can't stand the thought of physical exercise there may be another way to burn calories – thinking hard. Our brain uses 20–25 per cent of our daily energy resources – which is more than any other animal. Monkeys, for example, have much smaller and more economical brains relative to body size than us because they couldn't afford the luxury of such a gas-guzzling limousine. Apes would have to be eating for over twenty hours a day to get enough energy to feed a brain of our relative size. About two million years ago we made an evolutionary step change whereby our brains grew and our intestines shrank by a third – particularly our colons, which are now proportionally much smaller. The reason for this was cooking.

The simple idea of using fire to change the composition of plants and meat transformed us into modern humans. Suddenly, by using heat to break down the complex starches of root vegetables and leaves we could extract the energy and nutrients in a fraction of the time that it took before. We no longer needed to spend most of the day chewing food like cows do, and could risk going further afield and hunting. This also meant that we no longer needed to run our elaborate combustion engines – our very long large intestines – which were designed to give plenty of time to digest tough plants. Unlike apes we no longer depended on the energy (like short-chain fatty acids) released from food fermented by our microbes.

Reducing the size of our intestines enabled us to invest more energy and calories elsewhere – most obviously in our brains. The discovery of cooking and the ability to obtain calories easily are now seen as the major event that triggered our brain enlargement, leading to the emergence of modern humans and our subsequent dominance of the planet. Our large brains are greedy and consume about 300 calories a day, even when we aren't using them much. This is roughly equivalent to the energy of a weak light-bulb, and we can't turn it off – the energy we use up when we are asleep is nearly the same.

This supply of energy comes in mainly as glucose, and even when we are fasting or asleep the brain ensures it gets over half the body's supply of circulating glucose and so never goes hungry. Our brains are the greediest organ and use a fifth of our total resting energy despite constituting only 2 per cent of our body weight.[42] Just running our bodies at total rest costs us around 1,300 calories a day. The good news is that it is quite easy to expend energy. For example, just watching TV for an hour uses 60 calories; reading this chapter will expend over 80, and even more if you are on the chubby side or are finding the whole experience stressful.

We have seen how relying on counting calories to lose weight is often misleading and that trying to lose weight by exercise alone is futile. However, until we come up with a better system, calories are here to stay, and they give us at least a rough guide to the overall energy content of foods. The rest of the details on food labels show us the other macronutrient components of food that the industry and the government have agreed we can see. They were introduced so we could judge for ourselves which products are healthy and which we should be wary of. But just how reliable are the accompanying health messages that many of us have taken for granted?

I will continue to use the format of the classic food label – somewhat ironically, as these labels are over-simplistic and reductionist as well as misleading. All nutrients – by which I mean the tiny components of food that are critical for all the bodily processes – are important, and form a part of virtually all useful foods, which are complex mixtures of the different food groups.

3

Fats: Total

Eating too much fat is bad for us. This is logical. Fatty food intake causes a build-up of fat in our arteries, which clog up and give us heart attacks; and the same fat also builds up in our body and makes us fat. Cholesterol is traditionally seen as the villain here. It was the first measure of fat in the blood that could be assessed by doctors, so it has become synonymous with the risk of heart disease. This was an easy and clear story that doctors have told their patients since the 1980s – and they still are. Unfortunately, it *is* just a story: cholesterol has been wrongly framed as the arch-criminal it never was. Some fats in our diet are not only good for us, but essential.

Fat makes up around a third of our body weight and we can't survive without it. But the word and its use are both prejudicial and confusing. As well as being another common word for 'obese' or 'wide', and for the contents of a beer belly, it has a more scientific use. It is a term applied to any substance made of fatty acids, which take many different forms, most of which are essential building blocks for our cells and our life. The groups of fatty acids that together form fat are called lipids, a more precise term, and they are what I'm re-ferring to when I use the term 'fat' in relation to diet or the blood. Fats are not soluble in water or blood; they are mainly produced and packaged in the liver, and are transported around the body joined to proteins via the blood. Fats come in a variety of shapes and sizes to replenish the cells and provide energy for useful organs like the brain. We couldn't exist long without fat, and when deprived of it in the diet our livers will do whatever it takes to make some.

When lipids are joined to proteins they are called lipoproteins, and are much more useful and interesting than total cholesterol. They can now be measured accurately in the blood as high-density or low-density lipoproteins, called HDL and LDL respectively, and they transport cholesterol around. The low-density lipoproteins are the bad guys, allowing small drops of lipids to get caught in the blood vessel walls, which leads to a build-up of plaque and heart disease or stroke. If your liver produces a lot of HDL, the good guys, most of your lipids are transported safely to their target and disposed of without any collateral damage. When lipids are made up of short chains of fatty acids they are generally liquid (oils), and when in longer chains they are solid at room temperature (fats).

Cholesterol – a big fat mistake

The reason cholesterol is fairly useless as a medical marker (with a few exceptions) is because it's a mixture of both good and bad lipids, and this mixture varies between people. A high level of total cholesterol is usually a problem, as on average it contains more bad lipids than good. However, it is even less useful as a marker in women than in men, and in the elderly a high total cholesterol is strangely protective against heart disease. Increasingly, the ratio of two transporters of cholesterol around the body, HDL and LDL, is used as a marker of risk, although we cannot yet measure LDL directly. A much better marker of the amount of high-risk lipids in your system is another small cholesterol transporter protein called ApoB, which deposits cholesterol in the wrong places and opens channels in the vessels allowing the lipid to form the plaque that causes the damage. It is not, as previously believed, the total amount of circulating cholesterol that is important but where the cholesterol localises, and this varies widely. Most cardiologists use this more accurate blood test to assess risk, but it is more expensive to test and underused because of our obsession with total cholesterol.[1]

Fat is a key macronutrient in our diet and comes in many forms. Total fat is often the first thing you read on a food label, but it is pretty unhelpful as it could be either very good or very bad for you, depending on the type. Most foods contain a mixture of many different types of fats: the commonest are saturated fats, mono-unsaturated fats, polyunsaturated fats and trans fats. Within each of these categories there are many subtypes – for example, there are at least twenty-four types of saturated fat, usually lumped together on the label. Scientists have long thought we knew which combination of fats was good and which was bad for us, but the fact is that we don't really.

Moving in a gradient from what is traditionally considered good to what is probably bad for our health, omega 3 fats, a type of poly-unsaturated fat, traditionally come top. They are known as essential fatty acids and derive from our diet – mainly from fatty wild fish and from some plants like linseed (flaxseed) – and are probably benefi-cial for the heart in reducing lipids and inflammation (dampening down the body's reaction to threats of infection). They have also been widely promoted as good for most diseases known to man, including dementia, attention disorders and arthritis.

Confusingly, another very similar fat (omega 6) is also a poly-unsaturated fat found in most vegetable oils and nuts and can also come from fatty meat and some farmed fish fed on soy and corn. By contrast with its squeaky-clean omega 3 cousin it has developed a bad name where our hearts are concerned. Having a high omega 3 to omega 6 fat ratio in your diet was believed to be beneficial – a ver-dict based on reasonable, but mainly weak, observational evidence.[2] However, when supplements are given in randomised trials to alter our ratios, no clear benefits are seen. This lack of effect is confirmed by careful meta-analysis of observational studies, which also shows no definite or beneficial effect.[3] In fact in a large multi-country study of blood levels, high omega 6 fat levels were actually much more beneficial for the heart than omega 3.[4] So the hype over omega 3 supplements and the determination to foist fish oils onto us to the detriment of omega 6 is way over the top.

A 2015 study from New Zealand of thirty-two products from thirty-two countries found that fewer than 10 per cent had as much omega 3 as claimed, and most of the products had much less.[5] This mirrored earlier surveys conducted by the US, UK, Canada and South Africa.[6][7] You should also be wary of relying on these products as the majority of tested fish oil supplements don't contain the in-gredients they claim to. Nevertheless, both these fats are likely to be good for us – at least as foods.

Mono-unsaturated fats come mainly from olive oil and canola oil, derived from rapeseed. Although predominantly beneficial, the evidence is of variable quality, being much better for olive oil.

Polyunsaturated fats (sometimes called PUFA) come from natural vegetable oils and are fairly neutral or protective, but the claims that margarines containing them are heart-protective are exaggerated and not backed up by hard evidence.

Saturated fats come from animal meats and dairy products and are the traditional villains, depending on their origin. A subgroup of saturated fats is medium-chain triglycerides, which come from palm oil and coconut oil. These have been used extensively in countries like Sri Lanka and Samoa, who at over 25 per cent of total calories have the highest saturated-fat intakes in the world.[8] There is still a lack of good evidence for or against coconut oil's effects on health, notwithstanding its increasing hype and commercial promotion. This is mainly because it is unclear whether this particular type of saturated fat, medium-chain triglycerides, is beneficial or harmful. Many of the promotional websites claim there is a wealth of research on coconut, but most that I found were unscientific and some were plainly bogus. Some celebrity chefs now promote the health properties of coconut oil over olive oil, which is a worrying trend and totally lacking in evidence.

Trans fats (also called hydrogenated fats) are the worst kind, and as they are totally artificial come only from processed or fried foods. They were initially hailed as the healthy alternative to butter (we discuss them in detail later).

Cholesterol is singled out on food labels in the US and many other countries – below the other fats so that you can avoid it and its 'deadly' effects. But highlighting its content in foods is crazy, as proportionally there is nearly three times as much of the lipid cholesterol in 'healthy' foods like lobster, crab meat or fish oil as in 'unhealthy' lard, beef or pork. Eggs are packed with cholesterol, and many people stopped eating them decades ago because of erroneous advice to avoid cholesterol at any cost. Cholesterol is a complex lipid that is part of virtually every cell in our bodies: 80 per cent of it is synthesised naturally inside us and only around 20 per cent is eaten as food. As well as providing the protective and nourishing lining to the walls of our cells, cholesterol is a key ingredient of many vitamins and important hormones. It is only thanks to the unfortunate combination of there being an easy blood test for it and a rotten PR campaign that cholesterol has acquired such an ill-deserved reputation.

When did fat get a bad name?

The anti-fat campaign has its origins in many places and events, but it really began in America. One reason was the very public spectacle of the heart attack suffered by President Eisenhower in 1955 and his subsequent attempts at a healthy low-cholesterol diet. This diet failed to reduce his blood cholesterol or his heart attacks, which he later died from. The driving force of the anti-fat campaign was a Minnesotan epidemiologist called Ancel Keys who was famous for inventing the 'K-rations' for US troops in the Second World War. He spent a sabbatical in the UK and was unimpressed by the fattiness of the English diet at the time, which in his view largely consisted of greasy fish and chips wrapped in newspaper, bangers and mash, and eggs and bacon. He noticed that wealthier English men, like their US counterparts, who could afford the most food were starting to die from heart attacks, which had previously been quite rare. He returned to the US determined to get the funds to prove his hypothesis.

The key to his theory was the famous Seven Countries study which associated rates of heart disease in seven countries with different dietary fat intakes. The countries ranged from Japan, with virtually no heart disease, to England and the US, with plenty. Keys's correlations were very convincing and his conclusions clear: dietary fat levels equated to heart attack risk. He actually looked at twenty-two countries and the correlations were not all as convincing (or as well publicised). But no matter – diet was not easy to measure. The studies had a major impact on the press and on medical and public opinion. Policy was altered to reduce fat intakes.

Other observational studies confirmed the opinions of the anti-fat movement. A large population project, later named 'the China Study', amassed a huge amount of dietary data from sixty-five counties and 120 villages in rural China in the 1970s when it was still a poor country and bicycles were the main means of transport. They compared in detail the diets of each county collected a few years before with the current rates of over fifty diseases as well as a number of blood markers.[9] Dietary fat levels and blood cholesterol were half those of the US, and the most common diseases seen in the West such as heart disease, diabetes and cancer were virtually non-existent.

Colin Campbell and his team at Cornell University running the China Study believed that the lack of both animal proteins and dairy products packed with fat, plus the large amount of vegetables consumed, were the reason for the amazing absence of cancers or heart disease. The conclusion was that we should eat vegetables and give up meat and dairy completely. This gave vital evidential support to the growing vegan and vegetarian movements while comprehensively contradicting the Atkins high-protein movement. Campbell's book *The China Study* became a worldwide bestseller.[10] Bill Clinton is reported to have lost 20 lb on the diet after reading the book following his heart problems.

Early lipid researchers uncovered some rare families who on testing were found to have blood cholesterols at over twice the normal levels, and they often died of heart disease in young or middle age. These families were later revealed to suffer from a group of genetic diseases known as hypercholesterolaemia (more simply, high cholesterol in the blood) caused by faulty genes, and they were put on strict fat-free diets. These rare patients showed a clear correlation between their blood cholesterol and disease. If you lowered their cholesterol to normal levels via diet or drugs, their risk of dying decreased significantly. In the remaining 99 per cent of the population a high-saturated fat diet would slightly increase total cholesterol levels, and it was assumed this would also increase the risk of heart disease. Thus the image of cholesterol as universally bad was further cemented.

As the simple message, 'Fat is deadly', spread around the developed world our diets changed for the worse. As well as reducing the diversity of our food it deprived us of many nutrients. But as we have seen, food fats come in many shapes and forms, some good, some bad and some ugly. So before you routinely reach for the zero-fat labelled items on the shelf it would be a good idea to find out more about them.

4

Fats: Saturated

If eating saturated fat is so bad, why do the French, who every day eat much more of it than the Anglo-Saxons, suffer from less than a third the rate of heart disease of Brits and on average live four years longer than Americans? Nearly a third of French saturated-fat intakes comes from dairy products. Since the late 1980s, when epidemiologists noticed fourfold differences in mortality between the UK and France, the so-called French paradox has been a subject of much debate and speculation.[1]

For many years the British–French rivalry has extended from rugby matches, politics and trading insults to trading mortality statistics. Since records in France started to be accurately collected they have reported considerably fewer deaths from heart disease and a longer lifespan than the British. The French are proud of this, but many UK colleagues tell me that much of the difference is due to a reluctance to record deaths properly, with the same 'Anglo-Saxon rigour'. Others disagree, asserting that misclassification could only explain at most 20 per cent of the difference, and point to a consistent north–south difference across Europe. Even within France itself there is a wide north–south difference, which suggests that most of the variation between UK and France is due to the healthier habits of the southerners.

What do the French do differently to give them this amazing advantage? The list is long. The regular wine drinking, cheese or yoghurt with every meal, the long dinner conversations about politics, culture and food, the relaxed attitude to marital affairs, the thirty-five-hour week, spending the whole of August on the beach, their love of regular strikes and street demonstrations, or their high taxes on the super-rich? Or maybe it's simply that they appreciate their food more and enjoy the pleasure of savouring with family and friends the small tasty portions over several courses. Their choice of food is quite different too. They often eat raw meat, such as in steak tartare and steaks cooked rare and dripping with blood, earthy-smelling saus-ages made from intestines, unpasteurised cheeses, raw oysters and seafood, snails and frogs' legs. They also cook practically everything in garlic and butter or olive oil.

The French enjoy a regular diet full of living things. Cheese, wine and yoghurt all teem with living microbes that during fermentation help make the food they eat so tasty and stop it going mouldy. Red wine drinking as the explanation of the difference between the two countries in heart disease rates has become the most popular theory, and has done much to boost red wine sales in the UK and America, as we will see later.

Could high-fat cheese be healthy?

Meat and cheese are perhaps the two most popular high-saturated-fat foods. Let us look at cheese first. Everyone with high cholesterol is familiar with the doctor's advice to reduce or cut out cheese and to take a statin drug. Most cheese is made up of 30 to 40 per cent fat, and most of this is saturated fat which is traditionally considered the fat to avoid. The rest of the fat in cheese is of the poly- and mono-unsaturated varieties. Only about 1 per cent is actually cholesterol.

The French eat a hell of a lot of cheese – 24 kg (53 lb) a year per person – nearly double that of the average American and Brit (13 kg). Most cheese in France is eaten as real cheese purchased from a shop rather than as in the US, and to a lesser extent in the UK, in processed food products. These differences were even greater in the 1970s when US and UK total consumption was only a third of its current level. Charles de Gaulle famously asked in 1962, 'How can you govern a country with two hundred and forty-six different types of cheese?'

De Gaulle was uncharacteristically modest in underestimating his country's riches: France has probably double that number of differ-ent varieties now (the UK may have as many as 800), with many protected by law in the way they are traditionally made and with an Appellation d'Origine Contrôlée certificate of origin like the classi-fication for wine. Of the top ten bestsellers at least four cheeses are unpasteurised, which the French believe gives them more taste and special properties. There are twenty-seven words for describing the different tastes and the enormous complexity of cheese. It contains a wide variety of microbes including bacteria, yeasts and fungi, and hundreds of species plus thousands of known and unknown strains.

The more artisanal the cheese-making process, the less sterile the conditions and the more diverse the microbes that grow in and on the cheese. The hundreds of natural microbe species plus the yeasts and moulds, particularly on the rind, provide more taste and better textures than the more industrial preparations. Despite the fears of other countries, outbreaks of cheese-related food poisoning are very rare, except for some home-made Mexican cheeses, which are best avoided. The French have a large cheese-science industry backing up their global market and are starting to seriously research the role of microbes. Unsurprisingly, the French cheese-research centres report mainly positive news about French cheese.

Some human clinical trials have shown that cheese supplements could be used to maintain the microbiome in people taking anti-biotics, which normally knock out a large proportion of our healthy species. Unpasteurised hard cheese, when given with antibiotics, has been found to speed up recovery times and reduce bacterial resistance compared to sterile industrial cheeses. It was postulated that the cheese microbes may be helping maintain the greater diversity of microbes in our guts.[2]

When I visited friends in the Savoie region of France recently, the process of making traditional Alpage (high altitude) Comté cheese, which uses a recipe that hasn't changed for centuries, was explained to me. The explanation (over much wine and cheese) took an hour, but the basic process involves mixing cold and warm cow's milk outside in the spring mountain air (other cheeses use an added enzyme), which chemically cuts off the tail of the milk protein, which in turn allows it to curdle into lumps that stick to the fat. This lumpy mixture is passed through a fine linen net to drain away some of the liquid, then stored on old wooden shelves in a damp cellar. There the cheese is regularly rubbed with a milky rag from a vat on the cellar floor full of milk whey and brine, which gives the cheese a good crust teeming with microbes, including bacteria and fungi which change the acidity and the taste. For French cheeses the key to the many tastes are the other substances that the milky rag may be dipped in – such as, in the past, horse urine, which gave acidity as well as distinctive flavours.

Most real cheeses are left to age and mature (including hard cheeses like cheddar) and have a crust or rind that contains other, larger, microbes called cheese mites, visible under a powerful magnifying glass. These greedy creatures eat the microbes and cheese on the crust and make tiny holes to improve the flavour, but they are usually brushed off before the cheese goes to the shops. One cheese, Mimolette, used to arrive with so many mites crawling on it that the US health authorities banned it. After its banning, the bright orange cheese, a French seventeenth-century copy of aged Gouda, became a big black-market seller. Cheese fans liked the earthy taste of the rind. The mites, transparent and understandably chubby, can be seen wriggling happily in a hardcore YouTube video as they munch away on the cheese. The video comes with a warning that you may never want to eat French cheese again.[3]

These mites emphasise that cheese is very much alive – a living entity full of microbes – from the specialist milk bacteria, lactobacilli, to yeasts and fungi that are responsible for the tasty blue veins in cheeses such as Roquefort and Stilton. The US FDA (Food and Drug Administration) in their wisdom have also decided that having bacteria in cheese is a bit risky (unlike firearms), and have banned a number of other artisanal cheeses made from raw unpasteurised milk such as Comté, Reblochon and Beaufort. They have even recently announced a likely crackdown on cheeses that are allowed to mature on 'old-fashioned' wood surfaces that are difficult to sterilise. The perceived risk to the American public of eating traditional foods versus the 'healthy' industrial alternatives tells us lots about the balance of risk assessment in current health and dietary policy.

So while the safety-conscious FDA and the commercially minded US Department of Agriculture push ever more industrial cheese-type products of the sterile processed kind, containing little if any live bacteria, the French prefer their traditional cheeses. Even the ones sold in supermarkets are packed with trillions of microbes. If you leave some of these cheeses out of the fridge you can sometimes see them changing shape as the bacteria and yeast interact and battle with each other, breaking down the milk to produce energy. The high acid levels produced by the bacteria keep rival microbes away and stop the cheese going rancid.

The only signs of real cheese slowly going off are usually a mould that forms on the surface, like the famous penicillin strain or, in some cases, the strong smell of ammonia in the cheeses made with more moisture, like Taleggio, Limburger or Epoisse, which need eating more quickly. On their own the toxins produced by the fungi in cheeses can be harmful, but are broken down and safe inside the cheese. Whatever the smell, if the cheese tastes good, apparently the rule is that you can eat it safely.

I was particularly interested to see if the French paradox could be explained by the ingestion of huge amounts of friendly microbes contained in the cheese the French consume daily. I therefore conducted an intensive *fromage* experiment on myself (and four other willing volunteers from my lab). I wanted to test the best variety of French cheeses to provide a wide variety of microbes, so I asked an expert at my local cheese shop to advise me.

After a few days of discussion (and tastings) he came up with three unpasteurised soft cheeses for me: Brie de Meaux, blue-veined strong-tasting Roquefort, and a runny, smelly Epoisse that can be eaten with a spoon when ripe. I was to eat a large quantity – 180 grams a day (a normal generous serving is 30 grams). To help wash it down and to stick with French tradition I allowed myself two glasses of good full-bodied red wine, and for any hunger pangs three yoghurts a day. I usually have a bit of cheese once or twice a week, but I abstained the week before while I collected some stool samples for testing my normal levels before the three-day diet period.

For a cheese lover like me, this seemed a doddle. Day 1's breakfast was easy – a nice big slab of Brie de Meaux on some brown bread; lunchtime was my Roquefort on some crackers, with an apple to dilute the strong taste; and the evening meal was a salad and my delicious Epoisse with bread and wine – perfect. The next day's food was the same, and breakfast was easy; but lunch with the Roquefort was getting tough to digest – maybe because it is a whopping 31 per cent fat. In the evening the cheese I ate was still tasty but I started to feel quite full.

When Day 3 came I felt relieved that this was nearly the end of the experiment. From the morning onwards I had an odd bloated feeling, and as a by-product of ingesting much less fibre I ended up constipated for a few days. I felt full although my calorie count was not excessive. Each day from the cheese alone I was consuming about 800 calories and about 45 grams of saturated fat, way more than the 'recommended' allowance. And that was without counting my other foods and the yoghurt. I continued collecting stool samples for two weeks to see how long the effects of the cheese microbes would last.

Up till ten years ago, the only way to detect microbes was to grow them into visible colonies. You had to tease them to grow in culture plates for several weeks – and in those days we thought there was just a limited number of interesting bacteria. But it turned out that only about 1 per cent of our gut microbes are easy to grow in culture and these are the ones generally harmful to us – pathogens, in other words. New gene-sequencing methods have totally changed the process and uncovered the other 99 per cent of species we live with, most of which are never harmful.

I was keen to see the sequencing results when they came back from my collaborator Rob Knight and his lab in San Diego. From my samples they extracted the combined DNA from all the microbes, then using gene-sequencers measured just one gene that all bacteria have in common, called the 16S gene. Each different species of bacteria has a distinctive version of the 16S gene which gives off a unique individual signature. When the analysis was finished around 1,000 species were arranged in groups and sub-families, which could then be compared across different people. Our latest results are showing that the average British microbe profile is unhealthy and lacks diversity but, sadly, is still better than the contents of most American guts.

My baseline results were slightly surprising: the microbes from the stool samples coming from my gut looked slightly more like Venezuelans than most Americans. The two commonest groups (phyla) of gut microbes are Bacteroidetes and Firmicutes. I had a higher starting level of Firmicutes than I expected. One big question was whether any of the cheese microbes had survived the journey through my stomach and small intestine. It used to be thought that the stomach acid was so strong that it killed all microbes. Luckily for cheese microbes, this isn't true. After just a day of the cheese diet my gut microbes had started to change, with big increases particularly in a number of the lactic acid bacilli (lactobacilli) and in the yeast penicillium.

The effects of the lactic acid microbes lasted a few days after the cheese stopped, then things started to return to normal, suggesting that the microbes couldn't survive without more supplies. These results were reassuringly similar to a much more detailed experiment by a team at Harvard run by Peter Turnbaugh, who followed six volunteers on a meat and dairy diet (to be discussed later).[4] After two weeks the good news was that I had increased the diversity of my microbes by a small but significant amount. However, it turned out that the results of the other four volunteers' cheese diets could not have been predicted, and some didn't change at all.

Whatever our diet, our microbes always carry our personal signature. This experiment showed that the particular composition of microbes that we are hosting varies a lot and may be the reason why many of us react in different ways to the same food. After my supersize cheese experiment – it took about two weeks for my bowels to return completely to normal – I felt like eating cheese again, showing that, like a kid in a sweetshop, you can sometimes have too much of a good thing.

The saturated-fat scare has hit the rails

Heart scares about eating too much dairy were widely reported in the 1980s and 90s, and have persisted. Some of these were caused by animal experiments in which mice or rats were fed large amounts of saturated fat, which increased their blood lipid levels and gave them signs of heart disease. But mice and men differ in many ways especially related to diet and health. Other scares came from epidemiology and we now know that many of those early studies were flawed, particularly the observational ones.

There were many other possible explanations for the large national differences between heart disease rates, discussed earlier. Critics that were brave enough accused the powerful anti-fat guru Ancel Keys of being very selective in the countries and the data he used. Others came to the opposite conclusions with the same data.[5] Over the following years further studies were inconsistent and inconclusive. Nevertheless, the prevailing dietary-fat-leads-to-heart-disease hypothesis took root.

For many years the medical and scientific community who raised objections to this hypothesis were shouted down as crazy heretics. The invention and widespread use of the statin drugs reinforced opinions. These drugs, went the theory, in contrast to diet, rapidly reduce blood cholesterol levels and cut heart disease and mortality. Guidelines in the UK and the US suggest one in four adults should now be taking them. It was assumed this was due to statins' cholesterol-lowering effects, but it turns out that was a red herring. The drugs' main benefit comes from anti-inflammatory actions on the blood vessels, and they have both good and bad effects on many other diseases.[6] With fresh eyes we can now look back more objectively at the accumulated diet data. A 2015 study re-examined the six early trials of the 1970s and 80s and found that although diets could reduce levels of cholesterol – contrary to the conclusions at the time – they had no effect on reducing heart disease.[7] A meta-analysis summarised twenty-one large observational studies exploring saturated-fat consumption across the world comprising a total of 347,000 people. Of the 11,000 who later developed heart disease over the next twenty years, no association was found between amounts of saturated fat in the diet and subsequent heart disease or stroke.[8]

Now, the tide of evidence has started to shift in the opposite direction.

The ideal way to solve the problem of saturated fat would be to perform a gold-standard randomised clinical trial of a high-dairy versus low-dairy diet and watch the levels of heart disease. But this was thought to be neither ethical – a full-fat milk and cheese diet was 'too dangerous' – nor practical – it would need to last years and be very expensive. One compromise was to carry out six-week diet studies to explore changes in heart risk factors. One such study gave forty-nine volunteers an initial low-fat diet for six weeks, then added an extra 13 per cent of calories either as cheese or butter for another six weeks. The cheese group didn't increase blood lipid levels or cholesterol at all, whereas the butter group did, showing that not all saturated fat is the same.[9]

The results now seem clear, particularly if you separate cheese from butter consumption. Far from being a risk for heart disease, full-fat cheese (but not butter), despite the saturated-fat content, now shows not only no harmful effects but a consistent protective effect on heart disease and mortality.[10] So although we can't depend on the reliability of these observational epidemiological studies, which have misled us in the past, we now have a reasonable hypothesis that the regular eating of traditional cheeses could actually prevent some heart and other health problems, owing to the extra microbes. Highly processed cheese or boiled or grilled cheese contains few viable microbes and doesn't have the same benefits. Other products like milk and fermented products containing microbes may also offer some advantages, as we shall see.

As for explaining the French (or Mediterranean) paradox, cheese definitely plays a role, although no one will now be able to solve the mystery for sure. This is because death rates in both the UK and France, along with most of the Western world, have plummeted as a consequence of effective treatments that didn't exist thirty years ago. Although the numbers of people with heart disease are still high, we can now keep them alive much longer after a heart attack. This is mainly thanks to minor surgery to unblock arteries, to drugs keeping our blood thin and to our blood pressure remaining well controlled.

The cheese pizza diet

Dan Janssen is thirty-nine years old and comes from a small town in Maryland, famous as the place where Babe Ruth got married. He likes cheese as well as pizza. In fact he likes it so much he has eaten it every day for the last twenty-five years – for every meal.

He tolerates the tomato sauce but won't touch any other vegetable toppings. He washes it down with a sugary cola and usually eats a whole 14-inch pizza himself, which contains 45 grams of saturated fat and 1,300 calories. He clearly has an obsessive eating disorder but seems strangely normal in most other ways. He is slim, and apart from needing insulin for his diabetes, which he has had since he was a kid, he looks relatively healthy. His doctors have suggested he change to a healthier diet – but are amazed (and not a little annoyed) that his blood cholesterol and blood pressure are fine; and he controls his insulin with injections. He spent several years working in the local Domino's pizza restaurant before eventually deciding to start a woodworking business.

When people tease him or say 'You're going to die!' he retorts 'We're all going to die. But I'm going to die with pizza in my belly.' His fiancée Madeleine, who like him is a vegetarian, tries to get him to eat vegetables (technically, tomatoes are a fruit). He has tried to please her, but gags and can't finish the slice with even just a few vegetables on top. 'Why ruin a good pizza with toppings!' She has encouraged him to see a therapist, who believes his problems started as a kid.

'Like when I was four or five, we lived in the backwoods of North Carolina, where I went to day care in a lady's home. She would try to feed all of us Brunswick stew every day, which is not something you would ever feed a five-year-old. It's either chicken, pork, or rabbit with beef and okra, lima beans, corn, potatoes, and tomatoes. I would protest and try to run away, but she would grab me. I can't remember whether she would beat me or spank me, but I know that she would throw me in a closet as my punishment. I would sit in there crying and screaming for a couple hours until my mom came to pick me up.'

When asked if he had changed his habits after seeing his therapist regularly he said, 'No. In fact, one of the reasons I like to go see her is because she's in the city and I can go to Joe Squared [a pizza place nearby] after to get pizza.'

Exceptional stories like this are hard to explain via our conventional understanding of nutrition. Of course, we don't know if Dan will drop dead next year or live to be a hundred, but if the latter, we'd be surprised. His intakes of saturated fat are very high – way over most countries' official 'recommended levels' of 20 to 30 grams a day, and he eats very little fibre. But what if some people have actually adapted to this type of high-fat dairy diet and yet remain healthy? In the influential Seven Countries study after the war Ancel

Keys highlighted the island of Crete as the place where the lowest cholesterol levels (half of US levels at the time) and the lowest rates of heart disease were to be found.

Cretan cholesterol and centenarians

The Cretan villages were in mountainous areas, isolated and poor. Most of the people were shepherds or fishermen, and despite very hard lives and no real medical facilities there were large numbers of centenarians among them. What Keys and his colleagues didn't highlight at the time were the huge amounts of animal and vegetable fat and dairy produce these villagers were eating. A geneticist colleague of mine, Ele Zeggini, has been looking more closely at a sample of these villages fifty years later. It turns out they are all quite different, isolated from each other, and with strong local dialects and customs.

One small village that Keys didn't look at comprises just over four thousand people called Anogia (which means 'upper earth' in Greek). It is nearly 3,000 feet up the north face of Mount Idi, where they rarely eat fish but do eat a lot of goat's cheese and yoghurt every day. The only real difference in their diet over the last few hundred years is that they can now afford to eat meat (usually goat) more regularly, whereas before it was only for special occasions. They are also getting quite lazy and would rather drive 400 yards than walk it.

These villagers are part of a national nutritional study and have regular check-ups and blood tests. What they show is that their total cholesterol levels in their blood are higher (just over 5 mmol/l) and so in theory less healthy than in some places in Greece and similar to north Europeans'; but importantly, while they do get cancer, unlike the rest of the country they show no signs of heart disease.

What Ele and her team found was that a high proportion of the population carried a mutation in a gene called APOC3, which explained the extra-elevated levels of the good lipid transporter found in blood, HDL, plus low levels of the harmful lipid triglyceride, providing protection for the heart despite the high-fat diets. Now, this isolated and partly inbred village with plenty of cousins had something in common with another unexpected population the other side of the world who also consume vast amounts of cheese and dairy – the Amish in the US. Amazingly, they also share the same rare genetic mutation providing the heart-protection gene that usually occurs in less than 1 in 50,000 people.[11]

This story shows how populations might adapt to odd diets and environments over relatively short times – a bit like the Masai in East Africa with their high-fat intakes of meat and milk and their blood drinking, or nomadic Mongolians who exist on fermented milk, meat and little else.

As well as the genes changing, it could also be that the microbes adapt. Microbes can add a generation every thirty minutes and indeed, they adapt a lot faster than we can. I haven't yet been able to test Dan the Pizza Man, but he may well have some cheese-loving gene mutations, as well as definitely hosting cheese-loving microbes in his gut. Microbes do not yet appear on food labels, so it's hard to guess how many friendly microbes are still alive in the cheese toppings of a Domino's pizza. Apparently, they are made up of a blend of frozen cheeses and starches. However, unless you are really sure of your genes and microbes, I don't think I would recommend the cheese-pizza diet.

Industrial cheese is a by-product of the excess-milk lakes that built up as milk went out of fashion in the US and Europe. This was led by giant processed food companies like Kraft Foods, who in the 1950s developed methods for transporting cheese with a shelf life of several months around the US. They had hit products like Cheez Whiz, which coming as a bright-orange sauce was as far from artisanal European cheese as you can get. The process involved boiling and spinning the cheese or treating it with emulsifiers and many other chemicals so that the fat and the milk blended together, and with added preservatives it could last for months. This produced a sterile product (no live microbes) that could be, and still is, added to more or less anything to enhance taste, addictive qualities and consistency.

The combination it has worked best with is pizza, which has become arguably the world's most popular meal. It has steadily and frighteningly become the principal source (14 per cent) of the US saturated-fat intake and a third of total calories; and one-third of young Americans eat it daily. This is astonishing for a food that was only created in its modern form in Naples in 1889 (for Queen Margherita) and brought to the US in 1905. Most of the pizza eaten today is not, of course, the artisanal fresh product common in Italy, but cheap, processed and frozen in a business worth over $40 billion in the US alone. Some advertised pizzas have so much cheese stuffed into the topping and crust that just one slice equals 14 grams of fat and 340 calories. As long as it is cheap and durable, there seems to be no end to what cheese can be added to.

US cheese consumption has increased fourfold since 1970, ironically mirroring a decline in milk drinking because of fears of its fat content. Although it contradicts their own diet guidelines, the US Department of Agriculture (USDA) and the farmers are happy.[12] Exports of 'real' American cheese pizza have also boomed, especially in fast-expanding Mexico.

Fromage nature and mosquitoes

Another non-traditional form of cheese is made when you mix certain bacteria with milk. You can personalise these cheeses to uniquely suit you as an individual. All you have to do is take swabs of your armpits, belly button and between your toes and mix what you have garnered with milk, then add some lactic acid bacilli, and hey presto your own personal and very individual cheese! Christina Agapakis from Los Angeles (UCLA) created such works of art with Norwegian sensory artists for a recent exhibition in Dublin called *Selfmade*. The cheeses look like any regular cow's or sheep's cheese and each one is named after the donor of the bacteria. The bacteria used for making normal cheeses are closely related to those in the darker, less washed parts of our bodies.

The famously pungent Limburger cheese is made from the same bacteria that many people have between their toes (*Brevibacteria linens*), the ones that cause smelly feet. The composition of bacteria inhabiting these areas may make you more, or less, attractive to other animals. Mosquitoes in particular seem very sensitive, and different species will avoid some bacterial odours and make a beeline for others, explaining why some people appear immune to insect bites. Our lucky UK twins were tested recently in a lab study putting their hands into a plastic bubble full of mosquitoes, then counting the attempted bites. There were big differences, and the attractiveness of some people to mosquitoes was clearly shown to be genetic.

Smell is a very subjective sensation, as the UCLA team found out when they prompted subjects before a controlled smelling experiment. Those told beforehand that the bacteria smelled like cheese subsequently reported it as a nice smell, but others who were told it was taken from a body said it smelled disgusting. Christina has apparently tasted 'her belly-button cheese' and it was 'just like a normal mild cheese'. Human cheeses haven't yet become a regular part of our diet – but who knows? As the ultimate selfie, it could catch on.

Bulgarian health food

Yoghurts are another common source of our saturated fats, although the amounts vary widely across styles, particularly when you include the recently introduced low-fat varieties. Yoghurts are made from goat's, cow's or sheep's milk and come in many different forms, thicknesses and consistencies. Greek yoghurt, which is more solid as the liquid is strained off, is now very popular. Traditional Greek yoghurt also contains the most saturated fat, often 14 grams per pot, and it also has plenty of vitamin B12, folic acid and calcium. Basically, the more traditional and natural the mix, the more the amount of saturated fat. The popular low- or zero-fat yoghurts obviously contain the least, but these use either artificial sweeteners or several spoonfuls of sugar or its equivalent in concentrated sweet fruit to compensate for the lack of flavour, and usually contain fewer vitamins and nutrients.

In the early 1900s, the pioneering Russian immunologist Dr Elie Metchnikoff was working as the first serious yoghurt scientist. He was an eminent researcher who in 1908 shared (with the scientist Paul Ehrlich) the Nobel Prize for showing that white cells were not villains but actually good guys when it came to fighting infection. He suggested for the first time that bacteria, like white cells, had been misrepresented as always evil, and that a symbiotic relationship exists between them and us. He said that 'intestinal flora are the principal cause of the too short duration of our life, which flickers out before having reached its goal . . . It is hoped that the opening century will witness the solution to this great problem.'

He came up with his theory after observing that Bulgarian peasants ate a lot of local yoghurt and despite hardship managed to live relatively long lives. He proposed a novel idea for those days – one that seems quite obvious now – that a link exists between better health and longer life. His theory was that ageing is caused by rotting toxic bacteria in the gut and that consuming lactic acid (producing bacteria in milk and yoghurt) could counteract this, thus prolonging life. He took his own medicine and drank sour milk every day from then on. He outlived both his wives, who didn't share his tastes, and after working at the Institut Pasteur in Paris lived to the age of seventy-one.

One of his admirers was Isaac Carasso, a wealthy Jewish Catalan who was working in the Balkans just before the First World War when he heard of Metchnikoff's work and thought it had great potential. His firm became the global company Danone, now worth around 35 billion euros. Another disciple was the Japanese doctor Minoru Shirota, who in the 1920s in Kyoto was looking for remedies for preventing infection. He cultured special strains of so-called friendly lactobacteria, which he modestly named after himself (*Lactobacillus casei Shirota*). His commercial skills led to the worldwide marketing of the yoghurt brand Yakult in 1935. It's not clear how much yoghurt he drank himself, but he lived to be eighty-three.

Today, there could be an ecological downside to this surge in Greek yoghurt production around the world. The unused whey protein left over from the filtered yoghurt is so acidic that it is illegal to get rid of it in the normal way as it could devastate plants and wildlife. In the north-east US 150 million gallons of toxic whey form giant lakes as it waits to be disposed of. Eco-pioneers are experimenting with mixing it with animal manure, using the bacterial fermentation to produce methane gas, which although rather smelly could potentially allow yoghurt to power generators.

Despite the 'unhealthy' saturated fat and concentration of calories, dairy products could, paradoxically, help us lose weight. In several trials comparing dairy and non-dairy diets, the dairy group in each study lost slightly more weight. This was only significant if the groups were also calorie-restricting – that is, trying to lose weight at the same time – but the dairy consumers lost a significant amount of body fat while gaining lean muscle. This suggests that some constituents of dairy products could reduce visceral belly fat, which if true would be a major bonus. Although it is still unclear whether the actual saturated-fat levels are crucial or not, the passenger microbes may be.[13]

One of the recurring themes of this book is the surprising lack of good evidence for any health benefits or harm potential of any particular food product. Specifically for yoghurt eating and weight loss there have been only two randomised clinical trials, both small, short-term and inconclusive. There are, however, six large population (albeit observational) cohort studies following the yoghurt habits of over 150,000 people over time. Four of the six had a positive finding, such as a recent Spanish study of 8,000 men and women followed over six and a half years which found some slight weight loss when full-fat yoghurt was being consumed. This translated into a 40 per cent reduction in the risk of becoming obese if they ate at least one serving daily.[14] So again we see that increasing your proportion of calories from dairy certainly doesn't make you gain weight, as we previously believed,[15] and could even help a bit if you also want to lose a few pounds.

Short-term yoghurt experiments show improvements in the B vitamin thiamine, produced uniquely by human gut microbes. Other studies in mice with some special strains of lactobacillus (including the Bulgarian strain) have shown improvements in immunity.[16] But direct and consistent evidence of immune benefits in humans remains absent, apart from one small study which showed a reduction in common colds in the elderly. Our own twins studies are, however, showing some clear and important links in humans between microbes, diet and the immune system, which I will discuss later.

Super-microbes and probiotics

All yoghurt contains in large amounts the same group of bacteria that start the fermenting process in milk; the lactic acid bacilli, or lactobacilli, already referred to. These microbes help the digestion of lactose. Yoghurts vary in the amounts and the precise strains they contain, and in whichever other bacteria may be present naturally or artificially added. Most of the bacteria in natural yoghurt are strains that don't normally live in our intestines. When so-called gut-friendly bacteria are added to food in sufficient amounts and they are claimed to have potential health benefits, they are called 'probiotics'.

Probiotics are now big business. They come in several forms and are increasingly popular. As well as yoghurt and cheese, live microbes are contained in kefir, a sweet-sour fermented milk which is lightly carbonated. Kombucha is a slightly fizzy black tea which is fermented with both bacteria and yeast. Fermenting soybeans with fungi and salt produces miso paste, and when cabbage is fermented as sauerkraut or spicy Korean kimchi it can contain over a dozen probiotic microbes. Some can even be found in olives marinated in brine.

There has been much controversy about the health claims of probiotics added to yoghurts and other milk products. They are often sold separately freeze-dried in health food shops to combat antibiotic use or stomach upsets; others are said to increase and boost the immune systems of both healthy and diseased individuals, and they are marketed with an increasingly bizarre list of health claims. Most of the species sold commercially are related to lactobacilli or bifidobacteria.

The best scientific evidence that probiotics actually work comes not from yoghurt commercials but from research into the prevention of a nasty and potentially fatal disease known as antibiotic-induced disease, which often occurs in highly susceptible pre-term babies or elderly patients. Antibiotics (which we discuss again later) are widely used to treat infections when certain rare pathogenic bacteria species grow out of control. These often work, but they usually cause collateral damage, killing many friendly species and altering the natural community, the microbiome. This can lead to some of these pathogenic bacteria having no natural enemies and taking over; they become resistant to even the most powerful antibiotics.

A combination of lactobacilli and bifido microbes is added to many types of yoghurt, which are recommended to prevent the severe gut infection C. diff (full name, Clostridium difficile). This is a pathogenic bacteria that can take over the antibiotic-treated gut in a significant proportion of hospitalised people, particularly females and the elderly. A recent meta-analysis of twenty-one clinical trials found a 60 per cent benefit in taking these probiotics for three weeks; although it doesn't always work, on average one C. diff case was prevented for every eight prescriptions of probiotics, making it highly cost-effective.[17]

However, on the high street and the internet the market is not well controlled, and as well as their efficacy often being absurdly exaggerated many probiotic products contain contaminated or even dead bacteria, or sub-optimal numbers – this seems to be particularly true for freeze-dried supplements. This combination of factors has led authorities in Europe and the US to clamp down on the claims that the yoghurt companies can make without first conducting better trials. This has left the manufacturers in a catch-22 situation, as the authorities are now treating probiotics like new drugs, and if being tested for health benefits they have to show rigorous evidence of both efficacy and safety, which would cost millions. The companies argue that probiotics are just food and that the FDA don't demand a trial when a new type of porridge is released. For the moment neither side is giving ground, and large-scale responsible probiotic trials in humans seem a way off.

A meta-analysis of several probiotic trials concluded that there was little evidence of consistent benefit in humans, either because they don't work or because most are small and short-term studies.[18] One notable exception is the trial of a particular strain of a probiotic called *Lactobacillus reuteri* in people with hypercholesterolaemia (genetically very high blood cholesterol, mentioned earlier).

Could microbes be the new fat eaters?

There is an emerging link between microbes and our blood lipids. Germ-free mice suffer from high lipid and cholesterol levels because they lack a key product that accumulates in mice's gall bladders called bile salts. It turned out that microbes were crucial in enabling these bile salts to do their normal mopping-up job. A group of Montreal researchers tested patients with very high cholesterol levels first by feeding them the probiotic in yoghurt over two weeks – with favourable results.[19] They then put the same microbes into capsules and gave them to a similar group of patients over nine weeks. They found a 10 per cent reduction in all the harmful blood lipids and an increase in the protective ones which matched the beneficial change in the mice's bile salts.[20]

What is strange is that despite the increasing general evidence of the health benefits of yoghurt, there is still little evidence that the probiotic friendly bacteria actually survive and replicate in the human gut. Of the basic starter lactobacilli probiotic, studies show that only 1 per cent passes from the acid stomach to the duodenum and then the trail goes cold. Most studies can find no later trace of viable microbes in the stools, or evidence that they survived in the colon.[21] Increasing severalfold the concentration of microbes in the yoghurt or using a slightly different strain can sometimes pay off, but overall it looks like no one single microbial type suits us all. The common probiotic strains in commercial yoghurt may just work well in one person but not in another.

This may be because a particular gut has the 'wrong environment' owing to some individual chemical signals or conditions; or like a new kid at school, the microbe in question is vastly outnumbered, gets bullied by the other microbes and never settles. One small but careful yoghurt trial was performed, and with some interesting results. Seven young female identical twin pairs volunteered to take yoghurt containing five well-known species of bacteria commonly found in many commercial brands. They all ate the yoghurt twice daily for seven weeks. The US team, led by the micro biome pioneer Jeff Gordon, reassuringly found that fairly high levels of the five microbes reached the colons of the twins, particularly one of the bifidos which lasted for a week after the yoghurt stopped.[22]

However, against expectations these microbes were apparently not having much impact. The team found no difference to the composition of the other resident gut microbes, who seemed unfazed by the arrival of the newcomers. When they then put the same five microbes into a highly controlled group of mice, they got the same result. They could detect the yoghurt microbes but they were not perturbing the other residents. Some scientists would have stopped there, but the team did a series of complex tests to show that the yoghurt microbes had had major effects behind the scenes. They had somehow massively ramped up the activity levels of genes that control the breakdown of complex carbohydrates and sugars found in fruit and vegetables.

So the consumption of yoghurt microbes alters the way we break down other foods and initiate anti-inflammatory pathways. Our microbial communities work together in large interacting networks to metabolise our food in many different and complex ways. So introducing a single microbe may not be sufficient to alter the balance of the multitude of others, but it could still alter the overall metabolic balance of the community and affect our health.

There is a caveat. I worry that many commercial yoghurt brands are overselling their benefits, especially when they are adding only low doses of one or two of their special patented microbial strains. Many low-fat yoghurts are full of sugar (or puréed fruit), which could negate the benefits as sugar stops bacteria growing. Since yoghurt bacteria, like other probiotics, usually can't survive in most humans unless replenished, you may need to eat them every day to be effective. It could also be that the exact strain of bacteria or community of bacteria is crucial, and that we need to be matched up more precisely with our microbial soulmates.

Personalised microbes and designer yoghurts

Studies of probiotics have produced generally good, consistent results in lab mice that are genetically similar and have identical environments, but human studies are disappointing, showing little replication between studies. This could be because human guts possess an amazing diversity of microbial species between people. This poses the question, do our own human genes determine which microbes are attracted to us? This is important, as it might explain why some probiotics or yoghurts work for some and not for others.

Scientists like me with a genetics background often believe that genes are important for everything in our bodies that is biologically useful. Not everyone agrees. Non-geneticists say that the remarkable diversity between humans suggests that it is the random effects of our surroundings and of the food we eat that are the major determinant. Two earlier twins studies in the US produced no conclusive evidence of a genetic influence. But when I heard microbiome expert and future collaborator Ruth Ley present the results at a meeting I thought the studies were too small, and that we could answer the question properly with my cohort of 11,000 twins.

I had spent twenty-odd years performing twins studies on hundreds of different traits, from religious beliefs and sexual preferences to vitamin D and body-fat levels, in order to determine whether characteristics or diseases were mainly influenced by genes or by environment. The design is simple: you compare the similarity of identical twins with the similarity of non-identical or fraternal twins. If the similarity is greater in the identical pairs, then genes are definitely involved, as identical twins share all their genes and non-identical twins share only 50 per cent of them, like normal siblings. Using a bit of simple maths you can work out what proportion of the differences between people are due to their genes: this is called the per cent heritability.

Our faithful twins were persuaded to part with a tiny stool sample, which was kept frozen and sent to Ruth Ley at Cornell University. She and her team extracted the DNA and then sequenced one highly variable gene, 16S (mentioned earlier), that uniquely separates the species. By identifying the proportions of the thousand or so major microbe groups for each person we could start to compare them. The first thing we noticed was the lack of similarity between any two people – the variety was amazing. We may only share 10 to 20 per cent of our individual microbial species with each other, compared to sharing 99.9 per cent of our DNA. Our identical twins shared slightly more species than the fraternal twins.

It was only when we divided the microbes into the major classification groups, or phyla, that a much clearer pattern emerged. Although diet and other environmental effects predominated for many major groups, like the Bacteroidetes, many of the smaller subgroups of bacterial families, like lactobacillus and bifidos, found to have major effects on diet, obesity and disease, were also seen to be under genetic influence. This control is, however, partial and over 60 per cent of the effects on these microbes still come from the environment.[23]

Our results were a surprise to most scientists in the field. What this means is that the number and type of microbes (like lactobacilli) that grow best in our guts are in part controlled by our genes – it's a bit like certain flowers or shrubs preferring different types of soil. This could explain why the current limited range of probiotics work in some people and not others. Until a wider range of probiotics are developed and ways of delivering them to our colons are improved, real foods containing both a wide range of beneficial bacteria and microbe fertilisers would have a better chance of helping most of us than gambling on just a few added species of bacteria.

Working with a thousand of our twins and with Mario Roederer at the US National Institute of Health, we have just performed a novel experiment on the immune cells found both in our intestines and in our blood that communicate most with the microbes, called Tregs (short for T-regulatory cells). We found that they vary between people and are strongly controlled by genes.[24] So our genes determine to some extent which microbes live and prosper in our guts – using our soil analogy – and then, importantly, how our immune system reacts and responds to them.

So genes are not inflexible, as previously believed, but can be turned up or down like dimmer switches. This 'epigenetic process' is how we humans adapt to new surroundings and new diets. It is also how we communicate with our microbes and how our microbes can switch our genes on and off and manipulate us. In this way, over time our diets (or microbes) could slowly alter the way our genes fertilise (the soil of) the colon, allowing a greater variety of microbes to eventually settle and grow.

Listen to the brain in your gut

Microbes in our gut were essential in allowing us as newborn babies to develop a normal brain and nervous system. There is good evidence that microbes in general and lactobacilli and bifidos in yoghurt in particular can affect key centres in the brain via the brain–gut axis. The gut is the second-largest network of nerves outside of our heads and has been called our second brain. It has been estimated that the kilometres of neurones and nerve connections in our intestines are the same size as the brain of a cat – an animal we admire for its cunning, self-centredness and nine lives. We should listen more to our guts.

A complex system of signals relays information to and from the brain and the gut. These connections control many functions, particularly eating and digesting food, but we are finding they do much more: for instance, they can influence our mood. Sufferers and their doctors have long known that stomach upsets send signals to the brain to initiate nausea, prevent eating, reduce activity and dampen mood, which can lead to short-term depression.

We looked recently at rare twin pairs containing one depressed twin and one happy one. In the affected twin's blood we found altered levels of the key brain chemical serotonin. This chemical comes from our food except when we are fasting, when our gut microbes manufacture it for us. So changes in our microbes will alter this key brain chemical and, potentially, our mood. This could explain the odd sensations of a euphoric high reported by some people after fasting.

There are known to be at least sixteen specific messenger chemicals, called gut hormones, released from the intestines into the blood that send signals to the brain to eat more or less food. They are carefully regulated by our genes and by what we eat. In times of stress, the brain can also alter the function of the bowel via our emotions, which in turn can alter other hormones, leading in a vicious circle of symptoms to the disruption of microbiomes and even to depression. But these messenger hormones don't work alone.

We are learning that the immune system is another player in the crucial gut–brain connection. Cells like the Tregs, mentioned earlier, are in constant contact with both our microbes and our brains and act as messengers between the two.[25] In humans, one of the commonest complaints affecting our intestines is irritable bowel syndrome (IBS), which classically affects women more than men and has its peak incidence between the ages of thirty and sixty. The cause is unknown, although stress is a factor. What has always confused doctors and made them sometimes think the symptoms – and the disease – were invented consciously or subconsciously by the patients is the fact that 50 per cent of cases are associated with psychiatric symptoms such as anxiety or depression or chronic non-specific pain.

Stress, swimming pools and IBS

According to Sally, the worst thing about having IBS over the last twenty-seven years has been 'never being quite certain what to expect and how long it will last. It occurs before interviews, after shocks, like when my daughter was assaulted, or just when I feel very anxious. The bouts generally last from a few weeks to several months. I did have an investigation to see if it was cancerous. It's funny, though, because when I was growing up I had no idea that people "went to the loo" every day. My bowels used to go up to three weeks before I emptied them. And yes, it did used to hurt. Since having IBS I can go about five or six times a day when it's not great, when it equalises, and I now "go to the loo" most days.

'I did have salmonella at one point in 1989, but that didn't really affect them. My "trigger" seems just to be stress. My sister was raped and her boyfriend and my dad were beaten up and left for dead when I was seventeen, and my symptoms started a few weeks after that. I have tried various treatments, but they haven't helped much. I admit my diet is not very good – I eat a lot of white bread and go to Mc-Donald's more than I should. In fact a burger often triggers pain or bowel problems. I've not tried regular probiotics or healthy yoghurt because if anything they frighten me!'

Sally is overweight at 13½ stone and hasn't been able to lose weight. She puts her poor health and diet down to chronic stress and ten years of unemployment. We measured the composition of Sally's gut microbes compared to a thousand control women. Sally had less of the common Bacteroidetes species and, interestingly, seven times more Actinobacter than other women, which are more common on the skin. She also had much less microbial diversity overall.

IBS is a disease that was hardly recognised fifty years ago and now affects over 10 per cent of people in most surveys, but is especially hard to define as there are no specific tests. It is characterised by altered bowel habit, bloating and abdominal pain, switching be-tween bouts of constipation and diarrhoea, often ending in a dash to the toilet after meals. There have been over twenty gut microbe studies looking at groups of IBS patients. The results show clear ab-normalities in the IBS sufferers but no consistent pattern indicating exactly which microbes are altered; but all, like Sally, have reduced levels of microbial diversity.[26]

A few studies have tried antibiotics as a treatment, with limited success. Some companies are making antibiotics more efficient by means of a special capsule that protects them until they reach the small intestine and colon. Probiotics have had some partial success in treating IBS symptoms in most of the forty-plus studies, although many of them have been small, short-term and unreliable.[27] Up to half of IBS patients show some evidence of having experienced leaky intestines, when small amounts of chemicals or even microbes can cross from the gut into the blood. But whether this is directly related to the change in microbial diversity or happens after the disease has started is still unclear.

We all recognise that mood and eating habits often go together. Many human and rodent studies have shown that stress can lead to either losing weight or overeating and increasing blood lipids.

Only one group studied are immune to extreme stress and act cool under pressure. These 'Die Hard' Bruce Willis types are in fact germ-free mice that revert to normal wimps when bacteria are re-introduced into their guts. So, clearly, our microbes are crucial in transmitting anxiety. I remember crying when as a child I lost my mother at our local swimming pool; it is still a painful memory. Researchers can detect stress in rodents and have done the same to young mice. When, after separating them from their mothers, they forced them to swim laps, their microbiomes were disrupted and became less diverse. The anxiety and stress were consistently reversed by giving them lactobacilli probiotics, which suggests that yoghurt, not crisps, should be our treat after going swimming.[28]

Only a couple of good probiotic trials of mood in humans have been published. One showed women photos of angry and friendly faces and monitored their brain activity after four weeks of probiot-ics. They found no major changes in mood but did see a toned-down emotional response. Another, similar, trial involving both men and women showed a reduced blood cortisol level (which is increased by stress), also after a month of probiotics. One study, sponsored by a yoghurt company, showed that the microbes in yoghurt rather than the milk itself can light up key centres of the brain, again dampening down negative thoughts.[29] Ice-cream was earlier reported to have the same effect – and proved to be a marketing sensation for Häagen-Dazs. So let's not get too excited.[30]

Overall the story for the health benefits of probiotics is cautiously favourable. They clearly work well when the body is at its weakest and the microbes are disrupted or not yet formed, as in infants, after infections, or in the elderly. For the majority of normal healthy people (unlike lab mice), there is as yet no incontrovertible evidence from randomised clinical trials that for most of us taking them regularly provides any obvious benefit. But these are early days: we are testing only a handful of the protective microbes and we don't yet know what are the ideal conditions to provide them with.

A small number of classic microbes have been used by yoghurt companies for over seventy years, and with sales rising they don't want to change a winning formula. As discussed earlier, many commercial yoghurts while promoting their low-fat status contain lots of concentrated fruit, emulsifiers and sugar or sweeteners which could inhibit the growth or function of many microbes. So avoid these and keep your yoghurt natural and with plenty of microbes. Only a tiny number of them survive, so to boost your chances look for products that are likely to be genuine and with plenty of spare microbes, meaning over 5 billion colony-forming units (CFUs), hopefully listed in the small print on the label.

The overall conclusion here is that for most of us saturated fat is *not* the villain to be avoided at all costs. The saturated fat many people eat in products like cheese and yoghurt is not, as we have so often been told, unhealthy, but likely to be beneficial. This is provided the food is 'real' and contains living microbes, and is not over-processed or full of other unwanted chemicals and sweeteners.

5

Fats: Unsaturated

Jack Sprat could eat no fat,
His wife could eat no lean;
And so betwixt them both, you see,
They lick'd the platter clean.

So goes the nursery rhyme allegedly referencing the unfortunate King Charles I and his wife Henrietta Maria's tastes.[1] But which of them would have lived longer if there hadn't been a revolution and a beheading? This used to be an easy question to answer: most of us would have bet on fat-loving Mrs Sprat, aka the Queen, for an early grave. In the past fifty years, with changing views on fat, improved breeding, genetic testing and butchery techniques the levels of fat in all cuts of meat have dropped by up to 30 per cent in countries like the UK. But should we avoid these fatty cuts, or is this yet another myth?

After my health scare on the top of the mountain and my relatively lucky escape from more serious problems, I felt it was time to re-evaluate my lifestyle and see what I might change. I wanted to reduce my risk of future strokes and heart disease while still trying to enjoy a normal life for however long I had. I was desperate to come off, if possible, my two blood-pressure medications. I knew that exercise can reduce blood pressure, so I started doing more cycling at weekends, took up swimming and some gentle running round the park.

I thought then about improving my diet. My father having died suddenly of a heart attack at fifty-seven, since I share half his genes, put me in a high-risk category, and I wanted to try and avoid that genetic fate. My cardiologist tested my blood for other risk factors and said that my cholesterol levels at around 5 mmol/l indicated no cause for concern (the average levels in the UK are around

6 mmol/l). More importantly, I had a reasonably good blood lipid (LDL/HDL) profile. All the same, I should still try to reduce it further, which would further lower my risk of heart disease. He advised me also to reduce my salt intake. My penchant for salty nuts aside, I thought this would be fairly easy, but I knew that for many people salt reduction has only a minor effect on blood pressure, and I wanted to do more.

I decided I needed a more radical change – a diet makeover. I had reached a point in my life when I really wanted to know more about the food I was eating. After fifty-odd years of meat eating I would try going vegetarian – well, nearly. I would give up meat, but I didn't want to give up seafood, for a couple of reasons. I couldn't find anything scientifically wrong with it, and I was about to spend several months on my own on sabbatical in Barcelona surrounded by some of the best seafood in the world.

To improve my heart chances I thought I would also give up eggs and dairy (milk, cheese and yoghurt). I had briefly read *The China Study*, which suggests that all the fats and calories in dairy and even the proteins are bad for us. I thought about giving up alcohol, but only for a few seconds. I reassured myself that most of the studies showed a benefit for the heart of a mild intake, especially of red wine.

So I started my no-meat-or-dairy challenge. This was the first time I had attempted such a strict regime. To my surprise, avoiding meat was pretty easy, apart from the odd tempting tapas with *jamón ibérico*, and I managed to continue for several years – until getting down seriously to writing this book, at least. What I hadn't bargained for was how hard it was for me to give up cheese. Having a glass of Rioja without Manchego left a gaping hole in my pleasure centres, and the thought of never again tasting a nice Parmigiano or fresh Brie was too much to bear. My short-lived semi-vegan phase ended about six weeks later when I couldn't find anything to eat in an American airport – if it didn't have ham or turkey in it, it was covered in something bright orange they called cheese. I was so hungry that I gave in and bought a cheese pizza slice.

After four months of my pescatarian and (mostly) low-dairy experiment I had lost a bit of weight (about 4 kg) and felt virtuous and healthier. My raised blood pressure had come down a little (by about 10 per cent) and I was able to stop taking one of the tablets and reduce the other (amlodipine) to the lowest dose. My blood cholesterol had also gone down by 15 per cent to around 4.2 mmol/l, and my HDL/LDL ratio had improved. But were these benefits solely due to my low-fat and low-animal-protein diet, or to something else?

One interesting consequence of suddenly becoming vegetarian was having to deal politely with the baffled and sometimes outraged reactions of friends. More significantly, for the first time in my life I had to think carefully about what I put in my mouth. Those of us without major religious, cultural or health foibles are probably used to accepting most food put in front of us. At parties or functions I now found myself declining most of the appetisers offered, either because they had some real or processed meat or, just as often, because I couldn't make out the ingredients. The act of making decisions delayed the reactive eating impulse, and seemed to ensure that fewer unwanted calories ended up inside me.

By cutting out meat in general I was not just reducing unsaturated and saturated fats and protein in the meat itself, but by default I was cutting out most unhealthy processed foods that contain lots of added salt, sugar and fat. The other helpful side effect was that I was eating far more fruit and vegetables than ever before, and was discovering that I liked pulses and many other vegetables I had never heard of. I was becoming more adventurous when ordering food in restaurants, whereas in the past I might just have had a steak, chips and salad. As my diet changed and diversified I realised that making a small rule about, say, meat can have a major impact – if you can stick to it. As meat in some form is found in most savoury processed and frozen foods, I found I rarely used the microwave and was relying more on fresh food. Clearly, changing your diet is about much more than just excluding one arbitrary villain: it is about all the other compensatory changes you make.

Virtually all fatty foods are a mixture of unsaturated and saturated fat – technical terms that refer to the number of chemical bonds each fatty acid has to prevent it being saturated with hydrogen. Plant and vegetable oils like canola (rape seed) and olive oil, nuts and avocados, all contain large amounts of unsaturated fats. Meat is mainly made up of water (75 per cent) and protein, which we discuss in depth later, and a lot of fat, both saturated and unsaturated. Levels vary, of course, depending on the types of meat and the cuts.

In nearly all countries overall meat consumption has increased steadily since 1961, doubling in Europe by 2003 although still far behind US levels.[2] In developed countries although red meat eating predominates, increasingly people are switching to white meat (chicken and turkey), which usually has less fat. In the UK between 1991 and 2012 meat eating increased slightly, most of it due to a 37 per cent increase in poultry consumption.[3] The difference between white and red meat is more than skin-deep. Red meat is red because of the protein myoglobin in the specialised muscle fibres that are good for endurance, whereas chicken muscles lack myoglobin, which explains why you may see them making a quick dash across the road but not running marathons.

These sprinting chickens contain less total fat and the lowest proportion of saturated fat – two-thirds of their fat is unsaturated – compared to a roughly equal proportion of unsaturated and saturated fats in beef, pork and lamb. There is wide variation, depending on where on the animal the meat is cut from; some cuts of lean pork are similar to chicken, and minced beef and sausages can contain over 10 per cent saturated fat. The 20 per cent increase in deaths from heart disease associated with meat eating was thought to be due to the amount of cholesterol and saturated fat in meats that caused the disease in the first place. This ignores the large amount of unsaturated fat in meat, and as we saw earlier the most recent meta-analysis studies of saturated-fat contents in food have now cast major doubts over fat being the real cause. Importantly, the massive global experiment of replacing the total fat component of our diet with extra (mainly highly refined) carbs in recent decades has turned out to be a total health disaster.[4][5]

It is worth now taking a closer look at the different types of fat in meat and at what our ancestors ate.

Ancestral steaks

Hunter-gatherers around the world vary in the amount of meat they eat. In the tropics it is only 30 per cent, while the rest is made up of plants. We now tend to assume that eating lean meat is natural and healthy and that most people will sensibly discard the visible surrounding fat. However, our ancestors probably did exactly the opposite. Weston Price, an eccentric and energetic dentist practising in the US in the early 1900s, dedicated twenty-five years of his life to travelling the world to record the diets of isolated populations. He travelled from Alaska to Africa to find people untouched by 'modern degeneration', by which he was referring to diets and lifestyle. This was a concept that was hugely popular at the time and exemplified by vegetarians like lapsed Adventist John Harvey Kellogg, the inventor of Corn Flakes, who also promoted sexual abstinence, yoghurt enemas and other bizarre treatments.

What Price (and his long-suffering wife) found on his tour was a total absence of modern disease in tribes that ate traditional food. One characteristic of these primitive tribes was their love of the fatty cuts of meat and the vital organs – liver, kidneys, heart, as well as the intestines.[6] He even witnessed Native Americans giving away the lean game meat to their dogs – exactly the opposite to modern Western behaviour. Inhabiting a very different landscape, Inuits have few plant sources and so seek out the skin of whales and caribou liver, which when eaten raw contain their only source of vitamin C. In the past our bodies and our traditions recognised which edible animal parts held the key nutrients. We may prefer fatty foods for this evolutionary reason.

One important message here is that strong links undoubtedly exist between how you and your brain interact with food and how your microbiome functions. People that really enjoy their food may actually, via their brains, be able to both make themselves feel happier and stimulate their microbes. As noted in an earlier chapter, cultures like those of the French and the Mediterranean countries love their food and food traditions, and they still spend far more time eating and talking about food than the Anglo-Saxons. And their strong food culture tells them that whatever their grandmothers ate is generally good for them, and they continue learning to cook and prepare the same meals.

Unsurprisingly, they were relatively well protected by these food traditions from overreacting when, for example, 'expert advice' from the US on the evils of high-fat dairy foods emerged in the 1970s and 80s. These countries simply went on eating their healthy yoghurt and high-fat cheeses and meats. They are probably better off as a result, as they didn't replace the fat with refined carbohydrates. This contrasts sharply with most Americans and Brits, who with little or no common food culture reacted to the next wave of erroneous advice in a state of considerable stress. So not only do they make bad choices about the best diets for them and their families, by swapping real foods like meat and fresh cheese for margarine and processed-cheese pizza, but the associated stress and guilt may have adversely affected their microbes and their health.

Greasy foreign food

When the dictator General Franco in the late 1960s decided to open up parts of southern Spain to mass tourism, he relaxed the strict bikini laws and built hotels to attract Britons. The first tourists went on package holidays to Torremolinos on the sunny south coast and loved it. But they were shocked by the food they were served – 'chicken and chips floating in a lake of greasy oil or covered in garlic – revolting'.

By contrast they saw their own British food such as fish and chips fried in batter as healthy and wholesome, and couldn't wait to get back to their traditional full-English fry-up. The canny local Spanish hoteliers and their cooks soon adapted to the millions of British tourists arriving annually. Rising to the challenge, they provided 'English Breakfasts and Fish and Chips just like at home', with the addition of cold beer twenty-four hours a day but without too much visible 'unhealthy' olive oil.

Mediterranean countries have been making and using olive oil since at least 4000 BC, and it is part of the practice of most ancient religions. Spain produces 40 per cent of the world's olive oil, followed by Italy and Greece, but the Greeks are apparently still the biggest users, at 24 litres per person per year in 2010 compared to 14 in Spain and 13 in Italy. For Greeks to be drinking nearly half a litre a week seems quite hard to believe, unless they still use it for washing and on their hair. The cynical but more likely explanation is that they may be reselling it to the Italians on the black market.

In the UK its use has been increasing. In 1990 only 7 million litres were imported, whereas in 2014 the figure had risen to over 60 million, but outside trendy areas of the country this is still a drop in the ocean. Like Americans, Brits consume less than 1 litre per person per year – the same as the Greeks use in a fortnight. One tablespoon of olive oil contains plenty of energy: about 120 calories and 13 grams of fat, which should be very fattening.

For non-Mediterranean countries that were dictating to the rest of the world what a healthy diet should contain, the possibility that olive oil had some health benefits rather than being scorned as a cheap lubricant or hair oil came about only gradually. The idea of adding a few spoonfuls daily to the American diet was actually discreetly included in the US food pyramid in 1995, although most Americans took no notice.[7]

Ancel Keys and his original researchers were surprised that the Cretans not only had the lowest rates of heart disease besides Japan, but also scored much better than northern Greece in this department. The studies of Cretan fishermen revealed that in the 1950s and 1960s they were consuming huge quantities of olive oil, making up about 40 per cent of their daily calories, as well as using it in the form of cheap soap. The story went that they would even drink it neat for breakfast. When I asked some of my Greek colleagues they said they had never personally witnessed this but it was quite likely – for instance when, at dawn, the poor shepherd or fisherman wanted a quick cheap hit of calories to last him the day. I tried this for breakfast for a few days, but after experiencing a strange fainting episode, I wouldn't recommend it on an empty stomach.

The problem facing Keys's researchers was that a southern Greek diet was so different from a British or American one that it was hard to know what to pinpoint. They initially assumed that the benefit of the diet came merely from reducing the saturated-fat levels by eating much less meat and dairy than they did, and that olive oil was fairly neutral. As more became known about fats, the possibility that some could be beneficial as well as harmful slowly took hold.

Olive oil – the fatty miracle drink

Once again, we see how scientists seem driven to ignore the big picture and try to find the single magic ingredient that makes people healthy or sick. Dissecting diets and even simple foods into their important key components is a complex area that has generally eluded the nutrition community, and work done in this area has generally been misguided. Even something like pure olive oil is made up of several sub-fractions of different fatty acids that vary in length and structure. The main fatty acid is oleic acid, a mono-unsaturated fat with double bonds, and it constitutes up to 80 per cent of high-quality oil. However, olive oil also contains palmitic (saturated fat) and linoleic fatty acids (polyunsaturated fat) plus hundreds of other compounds including around thirty phenol compounds. Finding out which are crucial is not easy.

Olive oil's properties may also change with heating or frying, and proponents of other 'better' oils have criticised olive oil for having a low frying temperature and producing certain 'toxic chemicals' when overheated or burnt. This is another example of the field routinely picking on a few chemicals out of thousands and assigning some 'deadly' or 'health-giving' theoretical properties to them – usually way out of the natural context. In fact, there is no good data to show that cooking with olive oil is actually harmful or has different properties for health than when consumed cold.

Olive oil comes in three main types: expensive high-quality extra virgin olive oil, which has less than 0.8 per cent acidity, indicating its freshness and quality, plus a strong, sometimes slightly bitter taste. This comes from the rapid first pressing of the olive – which is performed at cold temperatures and marketed as 'cold-pressed'. Virgin olive oil is a grade lower, possessing higher acidity and still some taste; then finally there's regular olive oil. This last is made industrially by refining the leftovers and is cheap and normally tasteless, but may be given some flavour by mixing with a small amount of extra virgin oil. We now know there are huge differences between the health properties of the types, and most of the Italian and Greek oil produced now is extra virgin. Good-quality extra virgin olive oil contains the greatest amounts of the chemicals known as polyphenols, which have special properties and probably account for much of its health benefits. The low-grade oils used in spreads and processed foods probably have no equivalent advantages.

The Mediterranean diet: the proof of the pudding

With all the controversy over including fat in the diet and blood cholesterol, one fact remains. Namely, that people living in Mediterranean countries have consistently lower rates of heart disease and stroke than northern Europeans and Americans. This is primarily because of their diet and not some fluke of genetics. The Mediterranean diet has many current variants but for this discussion I refer to it as the traditional dietary pattern found in the olive-growing areas of Greece and southern Italy in the late 1950s and early 1960s.

The main characteristics of this diet are a high consumption of whole grains, legumes, other vegetables, nuts and fruits; a relatively high fat consumption (up to 40 per cent of total energy intake), mostly from mono-unsaturated fatty acids (MUFA, providing up to 20 per cent of energy) mainly from olive oil; moderate to high fish consumption; poultry and dairy products (usually as yoghurt or cheese) consumed in moderate amounts; low consumption of red meats, processed meats and other meat products; and moderate alcohol intake, usually in the form of red wine with meals.

As with our discussion of the French paradox, there are many possible explanations for the differences in the incidence of heart disease in the Med. It used to be thought that the sunshine just made people more cheerful and that this translated into less stress and a healthier heart. Sadly, the truth is somewhat different, and the people reporting to be happiest and most content are the Scandinavians: the pragmatic Danes, who in particular have low expectations in this regard, regularly come top of most polls. In fact, the sunny Mediterranean countries tend to be the most miserable and discontented, thus knocking that theory on the head. As we discussed, I believe regular consumption of traditional dairy products – cheese and yoghurt – is one important factor. The other key ingredient dividing northern and southern Europe is of course olive oil.

A group of Spanish researchers set up a unique and ambitious project called PREDIMED in the early 2000s to see if the many observational studies suggesting the Mediterranean diet was beneficial could be proven in a proper gold-standard clinical trial lasting years. The initial idea was 'to try to return these at-risk Spanish patients back to an average diet their parents would have had in the 1960s'. The researchers collected 7,500 volunteers from across Spain, all in their sixties and at high risk of heart disease. They allocated them randomly into one of three dietary groups, and all were given regular advice and support to help them stick to their diets.

The comparison diet group were to take a low-fat diet as recommended by most nutritionists, and advised to reduce the proportion of fat they ate as calories – which in Spain was already quite high, at close to 40 per cent. They were told to avoid meat, olive oil, nuts, snacks, dairy (unless low-fat) and to eat more fish, fruit, whole grains and vegetables. The control group were given extra (non-edible) kitchen accessories as incentives.

The Mediterranean diet groups, on the other hand, were told to eat more fish, vegetables and fruit but to continue dairy, white meat, nuts and olive oil and drink wine. These groups were further divided into two subgroups, one given 30 grams of extra mixed nuts to eat per day, and the other given an extra bottle of extra virgin olive oil each week to cook and eat with, to tempt them to consume four tablespoons a day. The study was originally designed to compare the rates of heart disease and diabetes between the groups and to run for ten years.

All went smoothly for four and a half years, until an independent committee stopped the trial. This had been set up for patient safety, to protect the volunteers from carrying on with an arm of the study that was clearly detrimental to health. The committee announced their findings in the *New England Journal of Medicine* in 2013. It was a knock-out blow for the high-fat diverse diet advocates and against the low-fat traditionalists.[8] Both Mediterranean diet groups with high fat consumption had 30 per cent fewer incidences of heart attacks, strokes, memory loss and breast cancer, as well as improved lipid and cholesterol levels and blood pressure. Although both Mediterranean diet groups did well versus those on low-fat diets, the

extra-olive-oil group did even better than the extra-nut group at preventing diabetes and in respect of several other parameters.

The diet study was not designed to make the high-heart-risk
participants lose weight, and we know that on average most sixty-
year-olds continue to gain some weight. But participating in any
kind of a trial is usually beneficial for health, and the low-fat group
receiving regular dietary advice and support gained only a kilo over
five years. The nut group did slightly better and lost a small fraction, but unexpectedly the olive-oil users lost over a kilo and, more
importantly, reduced their waist measurements, suggesting they had
lost more visceral fat.

Although the trial had directly tested only extra virgin olive
oil, they also recorded data on the use of cheaper varieties. These
low-quality products had no obvious beneficial effect on the heart or
diabetes risk, which explained some of the previous conflicting olive
oil results that didn't account for this quality factor.[9] [10]

Until recently the benefits of polyphenols in olive oil were thought
to come mainly from its antioxidant properties whereby they mop
up any excess cell-damaging chemicals and have a calming anti-
inflammatory effect. Other studies show that olive oil can somehow
switch off the genes (possibly by epigenetics) responsible for much of
the inflammation in the blood vessels that leads to heart disease.[11] But
research suggests a much greater role for our microbes. Over 80 per
cent of the fatty acids and nutrients from olive oil reaches the colon
before full digestion and comes into contact with our microbes. Here
the microbes feed on the rich mix of fatty acids and polyphenols and
break them down into smaller by-products, and at this point several
interesting things happen.

Some of the compounds produced act as antioxidants, then use
the polyphenols as fuel to produce a range of even smaller bits of
fat – short-chain fatty acids. These compounds are more interesting
than their name suggests, signalling to the body to lower harmful
lipid levels and telling the immune system what to do next. Polyphenols actively encourage some microbes to flourish, such as
lactobacilli that mop up and bind fat/lipid particles and clear them
from the blood. They also prevent unwanted microbes from colonising our guts. This reduces the incidence of infection from bugs

like E. coli that can cause diarrhoea, H. pylori that can cause stomach ulcers, and other bugs that cause pneumonia and tooth decay. Even the build-up of unhealthy atheromatous plaque in our arteries is due in part to abnormal and mysterious microbial activity in the damaged blood vessel – and polyphenols are likely to help reduce this too.[12]

The PREDIMED study was a major landmark in medical studies of diet, and the first to definitely show the beneficial effect of a sustainable diet on health. It clearly demonstrates that extra virgin olive oil and nuts taken regularly on top of a basic Mediterranean diet reduces the incidence of disease and early death. It is the only diet that has produced this high degree of evidence; observational or short-term studies only suggest changes in markers of risk. Most nuts have fat as their major component (49 per cent for almonds), which accounts for their calories. The types of fat vary, but only about 10 per cent is saturated and the rest poly- and mono-unsaturated. That the nut diet performed nearly as well as the olive-oil diet suggests they have similar mechanisms of action on our microbes. This makes sense because nuts, in addition to fat, contain a wide range of other nutrients such as proteins and fibre as well as polyphenols similar to those found in the extra virgin olive oil. (We discuss nuts in more detail later.)

In fact polyphenols are found in a number of other Mediterranean foods: in many brightly coloured vegetables and fruits such as berries, in cocoa nut, some green and black teas, turmeric and red wine. The fact that we are more attracted to these foods visually may have an evolutionary basis. Selecting healthy foods on the basis of colour (counting colories) rather than calories makes a lot of sense. Sadly, we still know little about how many of these polyphenols are biologically active or affect our gut microbes.

One study looked at different types of *sofrito* (the tomato-based sauce containing onion, garlic and olive oil that is used across the Mediterranean) and found, amazingly, at least forty different types of polyphenols.[13] It is likely that each ingredient on its own may not work particularly well without the others. Once you ferment these fruits and vegetables as pickles or alcohol, the number of polyphenols produced can increase exponentially. The universal use of extra virgin olive oil may be the crucial catalyst for bringing out the benefits of all of them.

Olive oil may have benefits over other oils because the complex juice comes from the whole fruit rather than just the seed, and this may be a general principle that applies more widely. It also means that its extraction is much simpler and doesn't require chemicals or solvents. So far its proven benefits extend to reducing heart disease and diabetes and possibly helping weight loss, but it has been claimed also to alleviate arthritis via its anti-inflammatory effects. More far-fetched are the assertions that it is a cure for baldness, increases testosterone and boosts male libido – Greek and Italian women may contest these claims, of course.

Olive oil is therefore the poster boy of the new diet order, dispelling for ever the myth that eating saturated fat is bad for you. Full-fat yoghurt and real cheese should come off the 'banned' foods list, unless you only eat it as frozen pizza. Picky Jack Sprat who ate no fat may well have died before nutrient-rich Mrs Sprat. Microbes are behind many of the health benefits, and offering us probiotics is a good first attempt by companies to give us as supplements a few helpful microbes. These few species don't work for everyone – we are all unique in this respect – but there can be no doubt that the diverse, real, fresh foods from the Mediterranean are what we should be trying to eat more of.

6

Trans Fats

Probably the most dangerous kind of food is one that is hidden, that doesn't appear on a label. Currently my favourite is Chinese Gutter Oil. This is the polar opposite of extra virgin olive oil. It was investigative journalists who uncovered the sordid practice by which recycled oil is resold for cooking. It is consumed by as many as 10 per cent of the Chinese (generally the poorest families and in street cafés). And it's produced by boiling and cleaned up by the addition of industrial chemicals.

It gets its name from being literally sucked out of drains and sewers, sieved to remove unpleasant solid sewer contents, then processed in home-made labs.[1] It is a lucrative and still thriving business, even though as well as containing known carcinogenic chemicals the oil is likely to increase the risk of heart disease and other illnesses. Last year a gang was arrested for supplying over three million litres of it to a hundred cities. They had achieved its extra flavour by adding fat from decaying animal carcasses. This toxic substance is giving modern Chinese cuisine a bad image and can't be good for the nation's poor gut microbes either.

A law making this business illegal was introduced in 2009, after the US complained that imported Chinese milk tasted odd. It contained melamine (a furniture resin). Other recent Chinese food incidents include chemically produced fake eggs with wax shells, walnuts shells hollowed out and their contents replaced with concrete 'nuts', steamed dumplings containing cardboard instead of meat, and chemically treated rat and fox meat passed off as beef.[2] McDonald's was one of many corporations involved in yet another major Chinese food scandal in 2014: their main supplier had recycled and reprocessed condemned pork, chicken and beef, some of which was over a year past the expiry date.

Of course, the Chinese didn't invent unhealthy chemical foods.

Food industrialisation started on a large scale in the ever-innovative US after the Second World War. The cross-country transportation of natural butter and lard for cooking was becoming expensive and wasteful because the products went off after a few days. This led to a drive to replace them with chemically manufactured vegetable substitutes that would improve the structural qualities of the products, increase their shelf life, and increase profits. When they first emerged they were seen as miracles of US ingenuity, but the first margarines were originally banned from having yellow dyes to warn consumers.

The dye became legal and a 'healthy'-looking cooking oil that was not only easily packaged and kept for months, but was also cheap, had to be good. Products like Crisco in the 1950s and 60s, made originally from leftover cotton-seed oil, were marketed heavily by Procter & Gamble. They had huge success, with recipe books and TV personalities encouraging housewives to use them for every meal.

The equivalent bestselling cooking oil in the UK, made by Unilever, was euphemistically called 'vegetable shortening' (Spry, Crisp 'n Dry) and promoted as a light and much healthier alternative to butter and lard. Unknown to the trusting public was the fact that to get these vegetable fat molecules to stick together required clever but drastic chemistry called hydrogenation, so as to artificially create new tough chemical bonds that were very hard for heat (or the enzymes of the body and its microbes) to break down.[3] The industry loved these versatile chemicals, which were used in a huge variety of processed foods and dairy substitutes.

Helped by the US-led obsession to cut down on natural fats, the hydrogenated-fat market boomed in the 1970s and 80s and its products were seen as the 'healthy' alternative to dairy. By the early 1990s, the FDA estimated that 95 per cent of all biscuits and 100 per cent of all crackers and most other snacks contained these trans fats, as they came to be known. Many Americans were consuming 10 per cent of their calorie intakes as trans fats from cakes, biscuits, pastries, burgers, ice-cream, chips and other fried foods.[4] A whole generation was brought up on the wonders of 'healthy' margarine and cooking oil. The first reports, in the 1980s, of adverse health effects of this chemical fiddling were largely ignored.[5]

It turned out that even small amounts – 1 to 2 per cent – of daily trans-fat intake were shown to massively increase lipid levels, and risk of heart disease and sudden death rose threefold, not counting the extra cancers. An estimated extra 250,000 Americans died each year just because they had consumed trans fats. But, hampered by the food lobby, no real action was taken for many more years.

In 2004 major global brands of snacks like Doritos and Cheetos still had considerable amounts of trans fats in them. In the US in 2003 a man in California successfully won a lawsuit against Nabisco, the biscuit company making Oreos, who subsequently withdrew trans fats. Only in 2010 did class actions start against Smucker Co., who promoted Crisco oil as healthy. The delays may have been coincidental, but for about fifteen years the world's largest food company, General Foods, was owned by the world's biggest tobacco company, R.J. Reynolds, who had considerable experience of dealing with health concerns and litigation.

Now, most Western countries have reduced or banned trans fats completely. While the US in 2015 only restricts trans fats to 4 per cent of fat intake (equivalent to around 1.5 per cent of total intake), by 2003 Denmark had already enforced a complete ban. For several years Scandinavian franchises of McDonald's and KFC had stopped using trans fats in vegetable oils for cooking chicken nuggets and French fries, while the US was prevaricating and protecting its food industry from any sudden change.

The UK introduced a voluntary limit and better labelling only after action from pressure groups in 2005, but still resists an outright ban, despite experts agreeing there is no known safe minimum amount. The medical community and the National Institute for Clinical Excellence (NICE) unsuccessfully called for a trans-fat ban and for the legal reduction of salt and saturated fats in processed foods, which would prevent 40,000 deaths a year. The situation has improved slowly. In 2010 the UK population was estimated to take in on average less than 1 per cent of energy from trans fats, and in the US it is still around 2 per cent. However, there remain big social and regional differences, so that some people eating cheap fried and processed foods are still taking in three times these dangerous amounts. Unfortunately and perhaps inevitably, the problem has been exported around the world.

In many developing countries, such as Pakistan, trans fats are still sold as cheap cooking oil (mainly fake traditional ghee), and its use accounts for 7 per cent of calorie intake and is a major contributor to the country's increasing rates of heart disease.[6] Trans fats can also be produced outside the factory, just by frying food in very hot fat; and, oddly, in cows' stomachs as part of the natural work of microbes. The small amounts of this otherwise toxic fat in cow's milk don't seem to be a major problem for us, though.

Different lactobacillus species, surprisingly, can produce tiny amounts of trans fats in our own intestines as well as potentially deal with tiny (but no larger) amounts of excess trans fats in our diet.[7] So if you can't resist the urge for an occasional blow-out on junk food potentially full of trans fats, choosing real cheese, yoghurt or pro- biotics for your dessert could possibly protect you.

One reason we think trans fats are so harmful is because of their effect on the small fatty acids produced from natural and artificial fats as signals for the healthy body. The fatty acids are key for effec- tive communication between the immune system, the microbes and the metabolism of fat. The disruption caused by creating these arti- ficial compounds leads to a major upheaval of our fatty acid signals and messes with our metabolism.

A bit off-colour

Jason, ten years old, enjoyed his crisps, so it was a little strange when one day at school during break he felt he didn't want to eat the large bag he always brought with him. He felt particularly tired and lethar- gic, had a throbbing headache and was nauseous and sweaty. He had found it hard to concentrate on his maths lesson, though that wasn't unusual recently. His teacher sent him to the school nurse who saw immediately that his legs had swollen to what seemed twice their normal size and his skin was a funny grey-yellow colour. He had never been slim, but his belly was protruding more than usual.

The nurse's anxiety increased when she found his blood pressure was raised. She phoned his parents but couldn't get hold of them, so took him straight away to the nearest big hospital: King's Col- lege Hospital in south London. Luckily, he was seen quickly in the specialist centre and the alert doctors recognised the signs of liver disease. His blood tests were frightening: his cholesterol and triglycerides were high and the liver-function tests were off the scale. He had fluid in his belly and legs from a combination of the liver failure and the pressure this put on his heart. His blood showed he also had Type 2 diabetes, which used to be called 'adult-onset' diabetes before it became common in children too.

Eventually his mum was traced and arrived at the hospital, where the doctors quizzed her. 'He has always been a bit overweight like me and has always been a good eater, though I've never managed to get him to eat much greens or vegetables – except chips. But I don't suppose they count. I guess his weight has probably increased quite a lot recently and he don't like playing football any more. He will get better won't he?'

Jason was weighed at the hospital: at 63 kg he was about double the normal weight for his age, although some of that was the excess fluid. He was put on drugs to bring his sugar level down and statins to control the excess fats in his blood. After two weeks without recovery and the results of an MRI scan and liver biopsy, showing the organ and the tissue around it were massively infiltrated with fat, it was obvious – his only chance of survival was a liver transplant.

Jason's story is becoming increasingly familiar to doctors, yet twenty years ago these cases were incredibly rare. Fatty liver used to occur only in lifelong alcoholics. Current estimates are that around 5 to 10 per cent of US kids now have blood tests showing fatty livers, the risk being higher in boys and in children with Asian or Hispanic genes. Most are overweight or obese, and all have low-nutritional-content high-fat diets. These are kids with a low threshold for storing excess fat. Their liver and their fat-storage cells get overwhelmed by the circulating fat and descend into a state of constant inflammation and stress. Liver transplant is relatively successful, but still one in three will die within five years.[8]

Junk food – the perfect obesity storm

That junk food is bad for you is not news – the combination of saturated fat, calories, sugar, chemicals and lack of fibre is an obvious signal. The lack of diversity in the diet, though, is an overlooked factor: as we have seen, 80 per cent of processed food is made up of just four ingredients – corn, wheat, soy and meat. Long-term studies consistently show that regular eating of junk food items like potato crisps, chips and processed meats leads to the greatest increases in weight compared to other foods.[9]

The most popular fast-food order in most countries is a Big Mac, fries and a large Coke, which in the US gives you instantly 1,360 calories and adds up to over half the average total intakes. Furthermore, much of it is fat, plus as a bonus the equivalent of nineteen extra teaspoons of sugar. Currently one in three Americans eat in a fast-food restaurant at least daily. Even in the UK, a third of children under ten now eat junk food every day. Fast-food culture has certainly changed our idea of the family meal, since the first TV dinner was invented in 1952. In the US nearly one in five meals eaten nationally is consumed in the car, and other countries are following the trend.

Since Ray Kroc took over from the McDonald brothers in 1948 and created the franchised fast-food empire, over 68 million people are now served daily in 118 countries. Like it or not, McDonald's has become a global emblem of American culture, and with its iconic golden-arch M sign is seen as both a symbol of cleanliness and efficiency and a target for animal-rights and health campaigners. Back in 1974 President Richard Nixon endorsed the Big Mac as 'the best burger in the US', second only to his wife's offerings. Maggie Thatcher in 1989 as the UK prime minister personally opened the new McDonald's UK HQ in her constituency of Finchley, endorsing their business model thus: 'You produce value-for-money food and added to that you also make a profit.' Other US corporations are also massively successful: Burger King, KFC, Taco Bell, Pizza Hut and Subway, for instance, with their huge global markets, all conquer hearts, minds and bellies.

Americans spent $6 billion on fast food in 1970, and $195 billion by 2014. Real food has trouble competing with the billion-dollar marketing budgets of fast and processed foods. Also, their prices have fallen in relative terms, helped by government subsidies on the four key ingredients, while the cost of fresh produce has increased over the last two decades. As the relative cost of eating out compared to cooking at home has dropped, so has the range of food alternatives. There is now a ratio of five fast-food restaurants for every US supermarket, another trend being mirrored around the world.

When the food industry began developing processed foods they were preoccupied with microbes and with keeping products on the shelves longer without them going off, particularly given the distribution problems of a country the size of America. They knew that fermented products like yoghurt or sauerkraut or pickles that contained bacteria kept products fresh, but cakes, biscuits and snacks were more of a problem. They worked out that if you added enough sugar it would inhibit bacterial growth, and increasing the fat content reduced the water content, which in turn reduced bacterial and fungal growth. Finally, on top of fat and sugar the third of the holy trinity, salt, was added, which also preserved the food and extended its shelf life. Together they would produce the conditions for the perfect obesity storm.

Products with all three of these Holy Trinity ingredients could last a very long time. When a Utah man looked in his old coat pocket and discovered a Big Mac in its wrapper that he had forgotten for fourteen years, he found it had no mould or fungus on it. Only the gherkin showed its age, the rest having dried out like a fossil.[10] Perhaps in future centuries such artefacts will be displayed in museums, like Tutankhamen's remains.

The companies found that when fat, sugar and salt were used in appropriate combinations, not only did they have a product that didn't go mouldy, they also had a winning combination for the public. Using high-tech labs and panels of tasters they could find the exact mix of the three that was irresistible for each item – what they called its 'bliss point'.[11] And once they had added the whole range of flavour enhancers and ingredients to change the texture, the poor consumer didn't stand a chance. It is no surprise that burgers, pizza, cakes and crisps (potato chips) all contain the big companies' holy trinity. Moreover, there is increasing evidence that we are adapting to seek this novel mixture of salt, sugar and fat, so different from real food.

Junk food can alter rat-brain activity in a manner some believe is similar to addictive drugs like cocaine. In one recent American study, after a diet of unlimited junk food (a good mix of highly processed bacon, sausage, cheesecake, pound cake, frosting and chocolate), the pleasure centre of some rats' brains became desensitised to the neuro-chemical dopamine after only five days.[12] This meant they required yet more of the same to keep the pleasure going.[13] When the junk food was stopped, the now obese rats preferred to slowly starve for two weeks rather than return to eating their healthy but less tasty replacement fare.[14] Evidently the effects of junk food on the pleasure-seeking centres in the brain persist way longer than the time it takes to eat a burger and fries.

Another study showed that a love of junk food in pregnant rats could be passed on to their offspring.[15] It could be transmitted to the babies by subtle (epigenetic) changes switching their genes up or down, or alternatively by maternal gut microbes that got passed on at birth, or via suckling. Some people claim to be addicted to junk food, but although they meet many of the criteria for addiction, whether addiction is really an issue here is controversial, as it clearly differs from addiction to artificial chemicals like glue or heroin. A similar debate addresses the question of whether a few rare people (usually celebrities) can really be addicted to sex, a pleasurable activ-ity that we are programmed to seek out.

The American documentary maker Morgan Spurlock famously went on a total McDonald's diet for thirty days for his documen-tary *Supersize Me*. His cholesterol went up 30 per cent, his uric acid levels (associated with gout) doubled, and the results of tests for liver damage more than trebled. He had gut ache, sweats and occasional nausea, and after a few days experienced strange cravings, depression and headaches, which were relieved temporarily on resuming eating. By the end of day 30 he had eaten 12 lb of fat, 30 lb of sugar and gained 7 per cent of body fat, most of it visceral. I was inspired to recreate the *Supersize Me* experiment, but with the aim of seeing how it would affect gut microbes.

Supersize Tom

I bravely considered doing the experiment myself, but after talking it through with my twenty-two-year-old son Tom it became obvious that he had far superior qualifications as a fast-food expert than I had. Tom, in common with other university students in the UK, has an appalling diet. Most students put on considerable weight that is often never lost; this is known in the US as the Freshman 15, after the average 15 lb (7 kg) gained. In term time Tom and his friends would regularly visit fast-food outlets like McDonald's once or twice a week (though, unusually for a student, Tom is a good cook). Anyway, the study appealed to him more than to me, and as a bonus he could write it up as his student project. The reason students eat so badly in the UK and the US is not solely financial or from laziness, but may relate to the stresses of living away from home plus the fact that cooking in many countries lacking strong food cultures is not seen as 'cool'.

We decided that ten days should be enough to see an effect without interfering with his work – or, more importantly, his social life. His only proviso was that he could alternate Chicken McNuggets with Big Macs. Although he had to keep his sugar levels high by having a regular Coke and a McFlurry ice-cream dessert (600 calories of sugar and saturated fat) with his main course, in the evening he could supplement the diet with vital extra nutrients supplied by crisps and beer.

His student friends were envious of Tom's sponsored junk-food fest. Students may get plenty of calories but not many nutrients, and I remember from my own time a fellow medical student who used the pub as his second home. He lived for two terms on nothing but cheese sandwiches and bitter ale, and developed bleeding gums and bruising. After several months, he was finally diagnosed with scurvy and given some oranges and lemons. So taking no chances, before he started I made sure Tom had a full week of fresh and varied fruits and vegetables.

His local restaurant was a fifteen-minute walk away, but for convenience there was a drive-through section for cars, which saved him valuable time and energy. He said there was no way he could eat breakfast there as well, but I suspected that was because he would have to get out of bed too early. So we agreed he could skip breakfast.

The first few days went well and he got to know his local staff on first-name terms and often brought along other student supporters. By day 3, though, the novelty had worn off and by day 4 he was starting to dread the second evening trip. By day 5 he was having fruit and salad cravings. On day 6 he noticed feeling bloated and sluggish after eating. On day 8 he had sweats after the meals which now left him tired for three hours afterwards; and he wasn't sleeping well. The last three days were the hardest with his tiredness increasing, and on day 9, unlike the addicted rats, he couldn't face returning for his evening nuggets and skipped the meal entirely. His productivity dropped and he found his assignments took him even longer than usual. Friends remarked that his skin seemed to have a yellow tinge and he looked unwell.

He was very relieved when the trial stopped. He had gained two kilos (4 lb) in weight and uncharacteristically went straight to the supermarket to buy salads and a fruit salad. It was six weeks before he could eat a Big Mac again, which he says is a record, and that was as part of a hangover cure. Clearly, although attracted to junk food, Tom wasn't addicted. Unlike Spurlock he didn't experience any cramps, cravings, headaches or vomiting – maybe because of his genes or because he'd already had plenty of training.

The results of Tom's microbiome tests after the ten days of burgers and nuggets were impressive. His levels of Bacteroidetes doubled from 25 per cent to 58, and his Firmicutes went down from 70 per cent to 38. His friendly bifido bacteria rates also halved. Importantly his microbial richness had been decimated and after only three days he had lost 40 per cent of his detectable species. Generally his microbes had taken on an aggressive and inflammatory fat-digesting function, particularly resistant to the extra bile acids his gall bladder was producing. His overall diversity profile moved in just a few days from being close to a healthy rural Venezuelan's that we mentioned earlier to looking like an average American's. Several rare species flourished on the diet, including one called Lautropia which is usually only noticed in immune-deficient patients. His profile was still abnormal with reduced diversity a week later, but then very slowly normalised.

A group in Harvard did a shorter but more detailed study on six of their lab volunteers who for three days ate a diet high in fat and protein made up of salami, meats, eggs and cheese with no carbs or fibre. What was special was that one volunteer was a lifelong vegetarian 'encouraged' to eat salami and burgers. His results showed even more dramatic changes that were seen within two days. Like Tom, he had big increases in Bacteroidetes and reductions in Firmicutes; and his Prevotella, which flourish in vegetarians, and normally reflect fibre in the diet, plummeted.[16]

These short-term but pretty extreme trials of junk food that sadly many people practise daily demonstrate the dramatic effects of losing nearly half your gut species. Such studies show that we can change our gut-microbe composition more easily than we had thought, just within a few days. The altered microbiome community produces a whole new set of metabolites and chemicals, which can alter our bodies in ways beyond just the effects of the fat and sugar. The reason that our microbes react to processed food so rapidly could be because they are literally being starved of fibre, while the brain is being over-loaded with fat and sugar. The other reason could be to do with the added chemicals and preservatives themselves. Although believed to be 'safe', emulsifiers have been shown to unbalance the microbes in mice and produce fattening pro-inflammatory chemicals.[17] Emulsifi-ers and binding agents are fairly universal in factory-produced foods like sauces and mayonnaise. Extrapolating from the mice, only 0.1 per cent of the food we ingest needs to contain these chemicals for them to have an effect on us. This really calls into question our food testing methods. The good news is that our microbes are flexible if kept healthy with diverse, high-fibre foods, and can reverse some of the unhealthy effects of these diets.

Toxic food and fast food mice

A group in Boston has been feeding mice the equivalent of Big Macs in liquidised form to see how they and their microbes fare compared to mice on normal diets. Both sets were allowed to eat as much as they wanted. Unsurprisingly the fast-food mice gained considerably more weight, particularly the dangerous visceral internal fat. As well as showing major differences in their microbiomes, the fattened-up mice were in a marked pro-inflammatory state, meaning that their cells were in alert mode as if being attacked, sending out signals to heighten chemical defences and make cell walls leaky.[18] Inflammatory states are normal in short bursts, but if sustained are unhealthy. Several other rodent studies have confirmed that high-fat and high-sugar diets produce these inflammatory changes as well as a leaky gut wall, which allows gut microbes and chemicals to pass through into the blood.[19]

The idea that the deadly mix of fat, sugar and salt plus a myriad of preservatives and chemicals that make up most processed food is itself pro-inflammatory has been around for a while, but without much hard evidence. However, a large number of rodent studies like the ones described above have now shown that when you feed them high-fat/high-sugar diets their bodies react as if under attack. Could it be that it is the food itself that makes the fat cells expand and produce the inflammatory signals?

Until recently it was thought that fat cells were boring storage depots with little contact with the rest of the body. We now know they are covered in helpful immune cells (Tregs) that help them communicate with the rest of the body's immune system.[20] Once someone becomes obese and the fat cells change, these response-dampening Treg cells on the fat cells' surface disappear, and the door is opened for all the inflammatory signals to be released. As we know, microbes and Tregs talk to each other, so could our microbes be playing a key role in making us fat?

If you feed germ-free mice these high-fat diets very little happens, and it needs the addition of microbes to make them put on weight, confirming that microbes are indeed crucial. In the Boston experiment and others, adding a probiotic like our friend lactobacillus or bifidos protected the mice from the effects of the junk food.[21] When the fat content of the diet increases rapidly there's a sudden increase in a certain type of bacteria that have a thick protective cell wall. Fragments of these cell walls (made up of lipids and sugars called LPS, or lipopolysaccharide) quickly build up and form an endotoxin, an internal poison that humans are very sensitive to.

We know LPS is the key player, because if you inject this LPS endotoxin into mice you get the same series of reactions as if they were eating the junk food, but without the brief fun. This includes triggering a reaction in the lining of the gut that starts the inflammatory process.[22] The gut lining then becomes leakier, allowing these toxic fragments into the bloodstream where they can reach the fat tissue and other organs like the liver. A chain reaction is started and the body goes into a state of high alert which we call sub-clinical inflammation, a biological version of a code orange alert.[23] A recent study of forty-five overweight and obese French subjects confirmed that regardless of body fat, junk-food diets with few vegetables lead to less microbial diversity and richness and more inflammation markers in the blood.[24]

What effect does all this low-level inflammation have on us? As well as sending increased stress signals to the body via the blood-stream which make our cells replicate faster and potentially shorten our lives, it affects the fat cells. These produce yet more inflammatory chemicals and signals which increase blood insulin, which in turn after a while prevents glucose being metabolised efficiently. This then signals for more (unneeded) fat to be stored, especially visceral fat around the belly. As we now know, this is not good.

Junk-food infections

In order to understand whether it is the microbes themselves that can actually make you fat or they are just a consequence of bad diets and the extra fat, some cleverer experiments were needed. Luckily, twins and germ-free mice again came to the rescue. From the local twin register Jeff Gordon's lab in St Louis found four American female twin pairs in their twenties (one identical and three fraternal pairs) 'discordant for obesity' (meaning one was obese and the other wasn't).

As expected, there were differences between their gut microbes, the leaner twin in each pair having a richer and healthier set with the usual higher rates of bifidos and lactobacilli and the fatter twin having a less diverse, inflammatory-looking profile. They then transferred the twins' eight stool samples and transplanted them into the germ-free mice, randomly dividing the lean and fat twins' output to see what would happen.

The results were surprisingly clear-cut, the mice receiving the fat twins' stool samples quickly became 16 per cent fatter, in particular putting on inflammatory internal (visceral) fat. This was clear proof that fat-associated microbes are really toxic and can be transmitted like an infection.[25] These toxic microbes are more likely to grow rapidly in our guts and be a problem if there is a community imbalance, if other microbes are suppressed, or if there is a general lack of diversity.

Up to this point the germ-free mice had been raised after their sterile C-section birth in isolated cages – in solitary confinement. To test whether the mice could make each other fat or thin by swapping microbes, the researchers now gave them a cell mate. Mice, like many other rodents, naturally eat their own droppings, but for a bit of variety they also like to sample their cell mates' too, and in this way exchange useful microbes. What the team found was a surprise.

The lean mouse, the one with the healthy microbes to start with, didn't get fat from the microbes of her obesity-prone neighbour. In fact, the reverse was true. The mouse with the toxic microbes was completely protected against obesity and inflammation by the transmission of the lean-mouse microbes (particularly the Bacteroidetes variety) into her guts. The other important finding was that this healthy transfer could be blocked by feeding them a high-fat/low-fibre diet and the mice still got fat. Conversely, a healthy low-fat/high-plant diet seemed to aid the healthy transfer, presumably making it hard for the original obese microbes to thrive. We performed a similar experiment with our twin study mentioned earlier. We transplanted a microbe associated with being lean into germ-free mice. This little-known microbe called Christensenella prevented mice from getting fat on high-fat diets.[26] The one in ten humans that have this microbe appear to be protected against obesity as well as accumulation of visceral fat, but unfortunately many do not.

This phenomenon of toxic-microbe infection could explain many observations in humans such as why fat mothers, even if not overeating, produce even fatter babies – who, remember, are much like germ-free mice. Changing the diets of pregnant women could potentially alter this vicious cycle. The data we have so far suggests people who are naturally lean or those on high-fibre diets seem relatively protected from the bad effects of the high-fat or junk-food diets. We don't know exactly why this is, but it is probably because they have microbes that are producing more of the healthy short-chain fatty acids (SCFAs), like butyrate, that keep the Treg cells happy and suppress the state of inflammation. Eventually, even these protective systems can be overcome, as after days of intensive high-fat, high-sugar junk food and no fibre.

Chinese gruel and rogue bacteria

Wu, living in Shanxi Province in China, had always been larger than his schoolmates. By the time he was eighteen he weighed 120 kg and by twenty-nine that had risen to 175 kg, and he had an impressive body mass index of 59. At only 172 centimetres (5 feet 7 inches) tall, he resembled a large barrel. Wu was not on any medication but he had a wealth of medical problems such as diabetes, high blood pressure, high cholesterol, abnormal liver-test results and very raised inflammatory markers. Basically, he was a mess. He didn't smoke, and only had the occasional alcoholic drink. He liked his noodles and his fatty meat and ate more than most Chinese of his age group, but not so much as to explain his extraordinary dimensions.

He was referred to Professor Liping Zhao in Shanghai, an expert in obesity and microbes. He suspected that Wu's case was rather special, and after routine tests to rule out other diseases he decided to look at the constituents of his gut microbiome.

A DNA sample of his stool revealed it was totally dominated by one type of bacteria called *Enterobacter*. This bacteria is usually harmless in small amounts in healthy people, but in this case was behaving like a ruthless rogue killer and producing a specific endo-toxin (B29) in huge quantities that attacked the cell walls of other competing gut bacteria. The invasive takeover had killed off most of his friendly species and also sent major aggressive inflammatory messages around the body.

Zhao put Wu on a specially designed diet. He was asked to eat around two-thirds of his normal calorific requirements, 1,500 calories a day made up of 70 per cent carbs, 17 per cent protein and 13 per cent fat. What was different was that it was composed of a combination of whole grains, traditional Chinese medicinal foods and foods that encourage the growth of the healthy microbes known as prebiotics. Amazingly this gruel-like diet worked fast.

Wu had lost 30 kg after nine weeks and 51 more after four months. These weight changes were mirrored in his blood, which returned to normal, as did his blood pressure. After nine weeks on the gruel diet, this dominating Enterobacter population and its toxins shrank from 30 per cent to less than 2 per cent and became undetectable after six months. This was mirrored by substantial improvement in the inflammation. As the toxic microbe was eliminated Wu noticed his feeling of constant hunger also subsided.[27]

Now, once again, it was hard to separate cause and effect. Had the state of obesity itself produced a weakening in the immune system that had allowed the dominance and strange behaviour of the microbe and its toxin? Or could the microbe itself have caused the obesity? Professor Zhao came up with a cunning plan to transplant the Enterobacter from Wu into the guts of germ-free mice. Now, as we have seen, these mice devoid of gut microbes don't get fat, even if overfed on high-fat diets.[28 29] However, when put on a high-fat/low-fibre (junk food) diet and implanted with just this one bacterial species they got very fat, and did so very quickly.

For the first few days all the mice lost weight, as a side effect of the inflammatory B29 toxin coming from the rogue bacteria. Then, within a week they were all putting on weight, and soon after started showing signs of diabetes, high lipid levels and signs of inflammation. Again, the combination of the high-fat diet and the toxic microbe was crucial, whereas the effects were minimal for mice fed on normal mouse diets.

Although this is only one isolated case it shows that a single microbe can directly cause obesity, just like an infection. It is probably (and hopefully) an unusual situation akin to a rare genetic mutation in humans. Other experiments using just a handful of implicated microbes have failed to produce the same dramatic effects. Usually to fatten up rodents the whole interacting microbial community is needed, and the same is likely true in humans.

After the end of the trial, a trimmer and more energetic Wu was so happy with the results that he stayed on his gruel diet for another year so as to lose even more weight. Prof Zhao was also happy with the outcome and he is now a big star in China, with TV appearances and a blog site that attracts over six million followers. Many more extremely obese Chinese of all ages are coming to him for his special diets.

Pang Ya was a three-year-old girl from north-west China who weighed 46 kg but whose parents were of normal size. The child had an uncontrollable appetite and would scream hysterically and employ many other tactics to get food. The desperate parents agreed to move to Shanghai close to the clinic and adhere to the strict Zhao diet for three years. The astonishing story of her weight loss and her regaining control of her normal microbes and appetite was the subject of a Chinese TV documentary.[30]

Over one thousand Chinese have so far been treated with his methods, many undergoing intensive investigation of their microbiomes. He has been reluctant to give away all his professional details but has published some early results of his special vegetarian diet.

In one study Zhao recruited ninety-three obese and early diabetic Shanghai volunteers who were placed on what he calls the WTP diet, based on a mixture of twelve whole grains, traditional Chinese medicinal foods and a selection of prebiotics. If you fancy whipping these up for breakfast instead of your Quaker Oats, the gruel contains a mixture of oats, adlay seeds, buckwheat, white bean, yellow corn, red bean, soybean, yam, big jujube, peanut and lotus seeds. Some patients get bitter melon too. The regime provides around 1,350 calories per day plus masses of fibre for nine weeks, followed by a maintenance diet for a total of five months. Most of the volunteers showed lowered levels of blood inflammation and insulin resistance, and the average person lost 5 kg while only 9 per cent failed to lose weight.[31] He has extended his studies and found that many referred children had undiagnosed genetic conditions that made them overeat (such as Prader-Willi syndrome). Nevertheless, these children also responded to the diet, which reduced – but didn't prevent – their excessive food cravings.

Keeping people on any diet for nearly six months is a major challenge, but when I spoke to Liping Zhao in London, he told me he thought he might have the answer. 'Compliance is a challenge in China like in other countries, but if you tell them the aim of the diet is to change their microbial community to reduce symptoms of hunger and other problems, this makes a big difference. It motivates them to continue treatment as if they had an infection. They are seen by a dietician weekly and every two weeks by medical staff, who at the same time analyse how the microbes are changing. As well as the special diets, patients are urged to complement their diet with vegetables, tofu and the less sweet fruits. Potatoes are forbidden.' His plan is to work out the top fifty microbe species found in obese patients and replace them with the top fifty found in lean people, effectively changing the keystone species to produce a healthy environment. Many of the Chinese herbs that Zhao uses have been around for centuries and have undergone plenty of trial-and-error testing.

Beware the golden juice

During the Dong Jin dynasty in the fourth century, a well-known traditional Chinese medicine doctor, Ge Hong, became famous for describing the first use of a special herbal mix as a cure for patients who had food poisoning or severe diarrhoea. This apparently produced great results, and several miraculous cures were reported in the first Chinese handbook of emergency medicine, *Zhou Hou Bei Ji Fang* (or 'Handy Therapy for Emergencies').[32] The exclusive top product was called Golden Juice, made from a selection of powerful herbs mixed with healthy human dung and clay, and left in a sealed urn buried underground to mature for twenty years before being served to the sick patient, usually in the form of tea.

It's a pity that Golden Juice was not on Zhao's list of twelve herbs, one of which, berberine, has been studied in great detail. Berberine is derived from the herb *Coptis chinensis*, which has been shown to prevent the inflammatory adverse effects of high-fat diets in rats and may act like a prebiotic, encouraging healthy microbes to grow.[33] A meta-analysis of a number of small-scale Chinese clinical studies suggests berberine could be used as an alternative treatment for diabetes, and it is widely touted as a miracle cure on the internet. Buyers, however, beware. Herbs like this often have strong effects and the quality and strength of the product is difficult to ascertain.

We come to the major problem of assessing the quality of Chinese clinical studies. Although at the top end Chinese scientists are now producing some of the best academic papers, at the lower end anyone can put out an academic publication if they pay for it. There is a whole industry – which even includes toll-free phone numbers – devoted to writing and publishing bogus papers for struggling scientists who have a few thousand dollars and want to succeed but lack the time, the ideas or, most importantly, any real data.[34] Sadly, the scandal of fake journals and studies is not limited to China: it is a global problem for science.

Liping Zhao has an interesting life story to tell. He grew up in a small farming town in Shanxi Province, and like most Chinese born on the eve of the Cultural Revolution, he and his two younger brothers had a simple upbringing. His father was a high school teacher and his mother worked in a textile factory. Both parents were firm believers in traditional remedies. Zhao remembers watching his father drinking a pungent, murky herbal concoction twice a day to try to fight a hepatitis B infection.

Zhao earned his PhD in molecular plant pathology and spent several years in the US at Cornell in the early 1990s learning about diet and health. He also learned first-hand that the American diet made his waist expand considerably. When he returned to Shanxi to start his laboratory, he focused on using beneficial bacteria to control plant infections. Throughout the 1990s he dabbled in research exploring the notion that bacterial strains might control infections in pigs, and might even work in humans. Meanwhile, his family's health was falling apart. His now overweight father's lipid levels had risen dramatically, and he had suffered two strokes. Zhao's two brothers had become obese as well. He decided to change his focus from plants and animals to human health.

After reading a 2004 paper by Jeff Gordon, the US microbiome pioneer who suggested that gut microbes could influence obesity, his interest was reignited. Lacking big funding, he used himself as a guinea-pig to try to pin down which microbes might be involved in weight gain. To Zhao, Western methods of weight loss that combined a low-calorie diet and strenuous exercise made no sense. 'Nutritionally, your body is under stress,' he says. 'Then you add to that physical stress. Maybe you can lose weight, but you might also damage your health.'

Having thought again about his father's herbal concoctions, he turned instead to traditional methods. He went on a diet of fermented prebiotic foods – Chinese yams and bitter melon – and whole grains to see if he could alter the bacterial ecology of his digestive system. During this personal experiment, he lost in just two years the extra 20 kg around his belly that he had gained in the US ten years before. His microbes became more diverse and one in particular, *F. praus-nitzii*, which has anti-inflammatory powers, increased dramatically. His results convinced him that harmful microbes could be replaced by healthy ones using special prebiotic diets. This motivated him to find the funds to start treating and investigating his now obese countrymen.

The great leap backwards

Many Chinese alive today remember the famines of the 1950s and 1960s caused by the zealous government movement for collectivisation, ironically called 'The Great Leap Forward', in which millions of Chinese died of starvation. We discussed earlier the 1980s studies' discovery that China had incredibly low levels of heart disease and cancer and their promotion of the Chinese diet as a potential saviour for the West.

A Chinese medical text written two thousand years ago entitled *Huangdi Neijing* identified obesity as a rare disease found among the elite class that was caused by eating too much 'fatty meats and polished grains'. You might well think that most modern Chinese have been elevated to this 'elite' status themselves, as the country now boasts more obese people than any other on earth. Looking at the changing shape of China is like watching a speeded-up film of the UK and the USA over the last three decades. More than a quarter of the adult Chinese population is overweight or obese and 7 per cent of children are clinically obese, indicating that the country's problem will worsen as they age. Many chubby children, lacking restraint when it comes to food and unwilling to exercise, are sent by desperate parents to US-style fat camps. This fat epidemic has already produced 100 million diabetics and 500 million pre-diabetics. The Chinese are even more prone, genetically, to diabetes than Europeans, laying down more visceral belly fat, and they are seeing a massive increase in heart disease. Although average calorie intakes haven't increased in the last decade, rising incomes have impacted on food consumption to the extent that meals now contain more than twice as much oil and meats as in the 1980s, when vegetables predominated.

Despite this increase in body fat, many kids in rural areas paradoxically suffer from vitamin deficiencies and reduced growth, so in common with other rapidly developing countries like India and many parts of Africa we are witnessing the weird phenomenon of both under- and over-nutrition going on at the same time. This is probably because when an individual has some persistent nutritional deficiency their body and fat cells experience an increased drive to lay down protective fat. These signals are likely to involve our microbes.

So the more poor-quality junk or processed food people eat, lacking in fibre and nutrients, the more signals the body sends to eat more in order to recover those 'missing factors', causing a vicious cycle of obesity and under-nutrition. And each country has its own version of junk foods, which are not just found in fast-food restaurants or cheap supermarkets.

Zhao believes the two main causes of the current Chinese epidemic are high levels of protein and fat (via meat) and a lack of whole grains, fibre and nutrients to build up the microbes and mop up the inflammation. A shift to dairy products is *not* the reason. When he was a child in the north, he remembers, the wheat noodles and rice were always greyish because they had been coarsely ground with plenty of fibre and nutrients. Now both the noodles and the rice are gleaming white, with no fibre or nutrients to attract microbes, and they keep for ever.

Many Chinese work long hours and don't have time to eat breakfast; they have a big lunch, usually provided free by their employer, then may go to dinner with clients or customers. Every dish contains meat, allowing little or no space for whole grains or vegetables. Now that women are working they have no time to cook, and most modern Chinese women no longer know how to prepare food or cook it themselves, leading to ever more reliance on restaurants or fast foods. There are over two thousand McDonald's outlets in China. Liping thinks the Chinese are now nutritionally more American than the Americans – and that's no compliment.

Microbe mind control and zombie burger eaters

If we know that these high-fat/high-sugar processed foods are bad for us, what drives us to keep eating them? What controls that inner desire?

It could be your microbes.[35] We already discussed how microbes via the brain chemicals they produce can influence moods, anxiety and stress. Each species of microbe has a preference for certain food sources, which allows them to feed and reproduce. They therefore have their own evolutionary drive to maintain their ecological niche and will do anything to ensure their survival. This includes sending signals to the hosting human that they want more of the same junk food that they thrive on.

This idea is now more than just a wild theory. It has actually been shown in mice that have been bred to artificially lack an immune receptor (TLR5). This disrupts normal gut–immune-system communications and leads to a change in gut microbes that trigger hunger. The hunger effects can be replicated by transplanting these microbes into normal mice and reversed by antibiotics – showing that the microbes are crucial.[36]

This mechanism has yet to be proven in humans, but it is likely to be particularly true of the toxic species that take over some individuals – the example of Enterobacter in Wu is a case in point. Small microbes controlling the larger host are well known in nature: for instance, the fungi that infiltrate the brains of ants and make them into 'zombies' to do their evil bidding. These ants are driven to climb up plants and then, feeding off the underside of leaves, they drop fungal spores onto uninfected ants underneath. Other bacteria can trick fruit flies into producing more insulin, thus laying down fat which helps the bacteria to multiply but doesn't help the poor fly.[37]

The notion that microbes could affect our food behaviour by providing us with brain-chemical rewards triggering us to eat more burgers is not so crazy – indeed, for our highly evolved and specialised bacteria it should be a piece of cake.[38]

We now see that our relationship with the fat in our diet is extremely complex, and that the simple dogma that we should all reduce total fat intakes has no scientific basis. Fats in processed foods with lots of salt and sugar are, we know, bad for us, and the artificially created trans fats are even worse. On the other hand, many fats such as the saturated varieties, previously condemned as unhealthy, turn out to be good for us as well as containing many key chemicals and nutrients that also feed and improve the diversity of our microbes. Fats in their multifarious forms are an essential part of many foods, so fixating on a few subtypes is futile. Furthermore, it diverts our attention from the importance of high-fat but healthy diets like the Mediterranean, in which diversity, colour and freshness are the key. So treat the 'zero fat' sticker as a sign of processing, not of health. Except in the artificial world of food labels, fat and protein are inseparable.

Let's now look at the effects of different forms of protein on our health.

7

Protein: Animal

William Banting was a highly successful west London undertaker whose famous family firm organised royal funerals. His health was good but he had always been on the heavy side. This had become an increasing problem in his thirties, and his waist had expanded way beyond all his peers'. Friends and dieticians had recommended different diets, and his doctor said he wasn't exercising enough. He was determined to lose weight. For the next three decades he tried different restrictive dietary regimes and many forms of exercise. For several years he tried rowing for one or two hours a day, but this just made him very hungry, so he turned to hours of swimming which didn't help either. He tried power walks, and steam-bath treatments in spas. But nothing worked and his weight remained the same.

Finally he consulted a private consultant, one Dr Harvey, an ENT surgeon in London who also dabbled in diet advice. He recommended a meat and fruit only diet, and after so many failures William was amazed by the results. He lost 29 kg in the next year and kept the weight off until he died, at eighty-one. His booklet *Letter on corpulence, Addressed to the public* was a controversial bestseller in 1864 Britain – a century before Atkins.

Most of the protein in our diets comes from just a few sources. These include meat like beef and chicken, which are over 30 per cent protein, fish like salmon and tuna at over 20 per cent, beans and nuts such as peanuts (24 per cent) and soybeans (12 per cent). Other artificially higher amounts come from soybean extracts and whey protein (from milk). Vegetarians and vegans can achieve a normal protein intake too, but they have to eat a greater volume of food. In a world where we raise and eat over 50 billion chickens a year, one of the big ongoing debates of our time is whether meat eating is good or bad for you.

Some social historians have called our switch from happy

hunter-gathering to oppressed agrarian cultures 'mankind's greatest mistake'.[1] Eating like our ancestors used to before farming started ten thousand years ago seems on the face of it logical. The Palaeolithic (paleo) diet, a variant of the amazingly successful Atkins high-protein diet, is hugely popular in the US, and there are thousands of paleo cookbooks available to prove it, as well as increasing numbers of trendy paleo restaurants (they make an exception for wine). Paleo meals contain plenty of animal protein and are low in carbs, without grains, cereals and most sugars. I met a woman from LA who said that she was ashamed to be the only member of her gym class not on a paleo diet (she couldn't bear giving up bread).

The proponents of high-protein/low-carb diets say you lose weight faster this way than by any other method, and keep it off. It supposedly also stops or reverses diabetes, reduces cholesterol and heart disease, and can cure allergies and auto-immune disease. Although rates of meat eating have declined in most Western countries, we are still highly dependent on animals for our calories, culture and family feasts. In the UK we eat an average of 84 kg per year, similar to our European neighbours but less than burger- and steak-loving Americans, who consume 127 kg annually. With meat plentiful and relatively cheap, should we all be embracing this 'back to our meat roots' organic philosophy?

The Atkins religion

Dickie looked into the bedroom mirror and didn't like what he saw. His rugby and squash playing days were over. He had to admit it: he was definitely podgy, he had a beer belly. His wife had been nagging him to do something for a while and she was right. His best days were behind him: a fifty-five-year-old surgeon, he was no longer in prime shape. He was always tired. He found being on his feet operating all day exhausting. Even playing golf was hard work, and his knee ached. He wanted to do something. He liked meat and liked the sound of the Atkins Diet. His brother had tried it and was still doing well after six months, having lost over 20 lb.

The first few days were easy; bacon and eggs to start the day, a couple of boiled eggs or a cheese omelette for lunch, and fish or steak plus salad for dinner. Once or twice he found himself, without thinking, going to grab a biscuit, a piece of bread or some grapes. After two weeks he felt good and noticed the difference in his waist-line; he was already shedding several pounds. Strangely, he didn't feel as hungry as he thought he would have.

He continued for another month and managed to lose a stone. He was proud of his willpower and his family were supportive. However, he started to notice a few annoying side effects: constipation was be-coming a problem, he was waking up with bad breath, and his initial burst of energy was waning. At the hospital he got a colleague to test his blood.

His blood lipid tests showed that his total cholesterol had increased about 5 per cent, which is within the error of the measurement, so wasn't very helpful. His unhealthy lipoprotein (LDL) had increased slightly, but reassuringly his levels of healthy HDL more than bal-anced it. More worryingly, his liver test results were slightly worse, as were his uric acid levels, a risk factor for gout. He phoned his brother to ask his advice, but he had not experienced any similar issues and was now slimmer and stable on a nearly zero-carb diet. Discouraged, Dickie began to ruminate on the idea of a nice piece of bread with his cheese and fruit. He soon relapsed and gradually slipped back into his old habits and, unfortunately, into his old trousers.

The Atkins Diet revolution started life in the 1970s as an alter-native to the many low-fat or low-GI diets of the time. Dr Atkins, although a hundred years after Banting, was going against the flow, and it took a while for his diet to become mainstream. But like a successful religion it would spread to millions of devoted followers.[2] When I suggested in my last book that Atkins himself was 'allegedly' overweight and had 'allegedly' had heart problems, I got more letters correcting me than on any other topic.

His idea, successful as it was, differed from most other diets of the era. It was simple and alluring, with no limits on quantity. This appealed to dieters who had been confused by complex regimes of mixing or alternating food types, by eating times and having to count calories and portion sizes. Just avoid carbs and have as much protein as you liked was the simple and effective message. Most people lost weight on the plan within a couple of weeks, and many sustained this for a few months.

The weight loss on an Atkins-style diet was often reported to be more rapid and noticeable than on low-fat regimes. Studies confirmed this was true for the first six months, though the evidence was less clear when the two diets were compared beyond a year. Trials also showed some benefits such as an increase in healthy HDL levels.[3] The combination of the high-protein and very low carb diet is also called a ketogenic diet, because the body deprived of glucose produces ketone bodies as fuel, made from groups of fatty acids in the liver stuck together. This is a less efficient fuel source, but a vital method of keeping our brains and other vital organs supplied with energy.[4]

The large metabolic changes in people on ketogenic diets cause side effects that can have strange advantages in treating some medical conditions, such as preventing seizures in children with epilepsy. High-protein diets may produce more weight loss short term than low-fat ones, as converting the essential energy from protein and fat is much less efficient than doing it from carbs: the body has to use more calories to do the same job. Another reason is that fat and protein provide a greater feeling of fullness (satiety) than most carbs, owing to the release of gut hormones that signal the brain. The high-protein and high-fat intakes may slightly reduce the laying down of fat, although this is controversial. And of course, the reduction in available food choices, as with many diets, may also reduce the total intake of calories.

For many people like Dickie, dieting for more than a few months is difficult. In fact fewer than one in six dieters said they have ever managed to maintain a 10 per cent weight loss for more than twelve months, and that is probably a big over-estimate.[5] This is due to boredom and lack of variety, and the other major explanation is metabolic.

Carefully controlled studies have shown that after a period of around six weeks of intensive dieting on any regime that has achieved a weight loss of over 10 per cent, energy expenditures and metabolism diminish so as to compensate as the body tries to regain its previous fat stores. This metabolic slow-down can be effectively as much as 10 per cent of daily calories. In trials, low-fat diets generally seem to provoke the greatest effect on this resetting mechanism, and high-protein/low-carb Atkins-style diets the least.[6]

But even high-protein diets can't fool the body for long. After a while, cortisol levels rise and thyroid levels fall, because both have the effect of increasing fat retention and reducing energy levels.[7] So although the mechanisms vary depending on the diet, the body always has a trick up its sleeve to ensure the replenishment of its fat stores. We all know people who have done well on some diets and not on others, and some who fail on most. This may not just be a case of willpower. Well-intentioned people may possess some factor that makes it more difficult for their bodies to respond to dieting. Most diets work for people in the first week or two, but much of the early weight loss is water. As the body reacts to lower calories longer term, major differences appear in how easily we burn off fat and how our metabolisms slow down to compensate. These mechanisms are complex, and regardless of whether they are due to gut or brain chemicals or psychological factors (as we learned in the introductory story of twins on diets) they are strongly influenced by genes as well as by microbes.

It is hard to distinguish in high-protein diets between how much the benefits and side effects can be put down to the lack of carbs and how much to the extra protein. During the recent evolution of the Atkins Diet the high priests of the billion-dollar corporation behind it have increasingly stressed the importance of low rather than zero carbs and reduced the emphasis on heavy meat eating. They have also tried to encourage the inclusion of more plant fibre in the diet. But despite more liberal regimes some people still have side effects like gout, constipation and bad breath. Strangely, mice have done badly too when put on Atkins-style diets for twenty-two weeks, which equates to several years for humans. They develop abnormal cholesterol and lipids, a pro-inflammatory state, increased liver fat, glucose intolerance and a shrinking of the pancreas, without even the merest weight loss to show for their trouble.[8]

Microbes predict your weight loss

The particular metabolic state of your body may influence your variable response to particular diets. My colleague Dusko Ehrlich was in charge of a large €20 million EU-funded microbiome project called Meta-Hit that answered some of the main questions. The researchers were able to look at gut microbes by sequencing the DNA of not just one gene (as with standard methods) but of every gene in every single microbe, and then putting them back together like a giant jigsaw puzzle. This is called shotgun meta-genomics because of the size of the data generated and the enormous scale of the computing problem in putting the pieces back together; and it costs several thousand euros per person.

They tested the microbe responses to diets. Forty-nine volunteers lost weight after six weeks on a special low-calorie (1,200) diet comprising 44 per cent healthy high-fibre carbs and high protein (35 per cent). They were then given another 20 per cent more calories for a further six weeks. They lost weight as expected in the first six weeks and then stabilised; some individuals then rapidly put on weight again. The prediction of who would lose the most was not based on their willpower or on their starting weight, but related to their gut contents.[9]

Although everyone benefited to some extent from this six-week low-calorie and high-protein diet, those subjects with the least rich and diverse community of microbes did the worst. They didn't manage to reduce the levels of inflammation in the blood and experienced the fastest rebound to their pre-diet weights. This low-diversity group made up 40 per cent of the French volunteers, and 23 per cent in a larger study of 292 Danish patients. The low-diversity group were on average more obese and had higher insulin and visceral-fat levels plus abnormal lipids, giving them an increased risk of diabetes and heart disease.[10]

The investigators found there were certain keystone species that were always present in the healthy subjects with rich and diverse microbe communities and usually sparse or absent in the unhealthy group. These included many of our friendly bacteria already discussed, like the bifidobacteria *F. prausnitzii*, lactobacillus, and the ancient methane-gas-producing bug (Methanobrevibacter). An analogy here is with keystone species in biodiverse habitats such as the case of the loss and subsequent reintroduction of wolves to the Yellowstone National Park. Without our keystone species the natural balance of the ecosystem falls apart.

The team found less inflammation and more of the beneficial fatty acid butyrate in the diverse group. They proposed that testing our guts for diversity (or richness) could be a new and better way to determine health as well as the future risk of many diseases like diabetes, and they are developing a clinical test.[11]

For the unhealthy subjects short of diverse microbes, the low-calorie/high-protein diet wasn't a total failure. After six weeks they did lose weight and managed to improve their microbiome diversity significantly. The problem was the rebound afterwards. The duration and intensity of the trial may not have been enough of a shake-up to permanently alter the microbe community. It wasn't clear from this study if it was the calorie restriction, the lack of carbs or the increased protein that had the major effects on the microbes.

Several studies have shown that the fewer fruits and vegetables and fibre you consume, the less diverse your microbiome, but the reverse is also likely to be true. Although I haven't tested it, one possible way to improve the outcome of an intensive low-carb/high-protein diet would be to start beforehand with a six-week fruit and vegetable feast to prime your microbes.

Reductionarians could save the planet

If weight loss is not your primary goal, what about the health effects of regularly eating meat? Vegetarians argue that meat is unnecessary for humans, causes animal suffering and adds to global warming. Studies have estimated that, because of the energy inefficiency of raising cattle with modern industrial methods, meat and dairy farming could contribute up to a fifth of greenhouse gasses globally, thereby contributing to climate change. This has led to calls for people to reduce consumption regardless of health or animal concerns and to become 'reductionarians' so as to save the planet. Although definitions are often movable, nearly 10 per cent of people in the UK now claim to be vegetarian or non-meat-eating, and the trend is increasing in many Western countries.

We surveyed 3,600 pairs of our British-born twins with a mean age of fifty-six in order to explore the causes of not eating meat. There were 104 pairs (9 per cent) of identical twins that were both vegetarian compared to only 55 (7 per cent) of non-identical pairs. This meant that although there was a small genetic component, it was outweighed by environmental and life-experience effects. These might include factors such as who they married, peer groups and where they lived. Vegetarians often like to quote studies of groups of vegans who avoid eggs, dairy and meat and live long and happy lives – but is this true?

Many Seventh Day Adventists (the protestant sect living in the US), who embrace healthy living, are vegans. When 34,000 of them were studied, researchers found that males, as well as being thinner, live on average seven years longer (females, four years) than average meat-eating Americans.[12] When the studies were extended to a further 70,000 Adventists across America they were able to compare, within the same group, the roughly equal proportions of meat-eating and non-meat-eating Adventists.

The vegetarian Adventists showed around 15 per cent reduced mortality (mainly from heart disease and cancer), but this more rigorous study translated into only about a two-year increase in longevity. This demonstrated the importance of controlling for other factors such as being Californian, being sporty, not drinking and being very religious.[13] These subjects believed that God wanted them to have the healthiest lifestyle possible, so they may also have been helped by divine forces? Several studies have shown that strong religious belief confers health benefits regardless of diet. Intriguingly, psychological studies of Dutch twins also show a connection between holding strong religious views and being unreliable responders to questionnaires. It wasn't that they deliberately lied, but rather that they had a tendency to tell people what they wanted to hear, which distorted their answers.[14]

The UK has over twice the proportion of vegetarians as the US, and the gap is increasing each year. This is in complete contrast to the numbers practising religion, where the steady threefold greater numbers in the US are outstripping the diminishing UK numbers. This may not be just coincidence. Our twins studies have shown belief in God has a partially genetic basis, as does a tendency to follow strict dietary patterns like veganism. Vegetarianism in many parts of the world started as part of religious movements, such as Hinduism, often as a way to distinguish them from other religious groups.

The health benefits of their dietary regime to over 30,000 British vegetarians and pescatarians (fish eaters) have been shown to be less clear cut than in Adventists, and it has been difficult to distinguish the effects of the lack of meat from those of a greater health consciousness. Although most studies show a reduction in cancers (up to 40 per cent in the latest fifteen-year follow-up) and 20 per cent in heart disease, this is balanced by increases in other diseases like strokes and little or no reductions in total mortality.[15][16] There is also a suggestion that British vegetarians are somehow less healthy than vegetarian Americans, which may be due to differences in culture, lifestyle, lack of religious beliefs, or to other not so healthy components of the British vegetarian diet such as baked beans, crisps or extra sugar.

Identical-twin studies are a great way to adjust for cultural and genetic factors and explore meat eating without many of the biases of observational studies. We looked closely in our TwinsUK study at our 122 British identical twin pairs who differed in their meat-eating habits, one being vegetarian or vegan and the other a meat eater. Remarkably, there was only a small difference in obesity within the twins as measured by BMI. The vegetarian was slightly slimmer by an average of just 1.3 kg (although the range went up to a 40-kg difference in one pair). This compares with much larger differences of about 4–5 kg in the Adventist studies, showing the significant effects of genes and culture which are hard to account for in non-twin studies.

Intriguingly, in our study we found that even if you were a regular meat eater, having a sister who was vegetarian made you healthier than the average UK twin, in that you'd be slimmer and less likely to smoke. Although we haven't accounted for the amount of meat eaten in the discordant twins, what is clear is that the differences in weight due purely to avoiding meat are exaggerated when you don't account for genes and upbringing.

Meat eaters and paleo-diet converts point to the facts of our evolutionary past as a solid basis for a dietary theory. There is no doubt we are omnivores, with bodies and digestive systems built to eat a variety of foods, both vegetables and meat. Our jawbones and teeth are made for chewing tough foods, and although the process is helped by cooking, we are different from fruit-eating primates in this respect. We also have an armoury of hormones and enzymes to break down proteins – not forgetting our ever-helpful microbes.

One big argument against totally excluding meat from our diets would be the lack of other easily available nutrients. Many vegans and some vegetarians run into nutritional problems as meat contains many essential nutrients like vitamin B12, zinc and iron that are very hard to find in vegetables. Vitamin B12 deficiency is very common in non-meat eaters and it is possible that this offsets some of the advantages of the vegetarian diet.

The British are still known by their stereotype as big strong meat eaters – *les rosbifs* as the French ironically call us (and we sometimes choose to call them 'Frogs'). One theory why British cuisine is traditionally so bland is to do with the alleged high quality of our meats over the last centuries, due to our fertile soil and wet grass. This is contrasted to the scrawny animals in France and Italy that have needed inventive and flavoursome sauces to mask their bad taste and texture. Yet in 2015 there are four to five times more vegetarians in the UK than in France, although the French might argue that our overcooking of dull meat could account for this.

Vegan vitamins

I mentioned earlier that I had a brief moment of veganism. This trial only lasted about six weeks, as I found life too tough without cheese and my fine dining was hampered when I was travelling abroad. But giving up meat was not a problem as long as I could eat fish. I happily did this for a year until I went for a medical check-up and noticed my blood levels of vitamin B12 and folate were low and my homocysteine level – a marker of heart disease risk – was high. I was ingesting plenty of folate from my vegetables, but the key vitamin B12 from meat was lacking and this lack was blocking my absorption of folate too.

This was annoying, as I had lost a few kilos and was feeling good; but my blood pressure was slightly raised and the low B12 was possibly making things worse. I started taking increasingly large amounts of B12 supplements each morning, but they had little effect on my blood levels. I tried eating a few eggs a week as they have some B12, but that didn't work either. Finally in desperation I tried B12 injections in my bottom. They worked, and my B12 and then my homocysteine level too finally returned towards normal. A few months later, as I was about to get myself (via instructions to my wife) another shot in the bottom, a thought came to me. This is daft. I am trying to be fit and healthy, yet having injections every month feels neither healthy nor particularly natural.

I decided I should eat just one steak a month and see what happened. So that's what I did – either a rare steak or a raw French-style steak tartare, once or twice a month, did the trick, and gave me the vitamins I needed without any artificial supplements. And the little experiment brought home to me not only that my body wasn't adapted to change so rapidly to a meat-free diet but that even my microbes couldn't manufacture all the necessary nutrients.

So if a little meat was good for me, was this a quirk or down to evolution?

The world according to paleo

The full paleo-diet doctrine prohibits grains, legumes (which includes peanuts), milk, cheese, refined carbs, sugars, alcohol and coffee. It also bans tomatoes, potatoes, and aubergines as they are all nightshade plants and thought to cause auto-immune disease via leaky guts. The diet encourages organic grass-fed meats and poultry, fish, coconut and olive oil and other vegetables and a small amount of fruits, though some adherents eat only berries. Like most religions paleo belief comes in different degrees of orthodoxy and strictness. The diet is broadly based on what we think was consumed for a million years or so before our very recent history, items for which we are perfectly adapted.[17] The main theory, the rationale (which is common to other diets such as those that avoid grains), is that our bodies haven't had time to evolve or adapt. But it has major flaws.

In my view the theory's chief shortcoming is that it doesn't account for the latest genetic or evolutionary studies and treats humans as rigid unchanging automatons. It also forgets about the trillions of microbes we carry around with us that have also been adapting and evolving. And can we, anyway, be certain what our ancestors actually ate? Was it the lean steaks and rocket salad that gym fanatics in LA imagine? As our ancestors didn't leave us any recipe books or DVDs, the situation calls for a good deal of speculation, relying on the observation of a few remaining hunter-gatherer tribes and archaeological remains, including bones, and the examination of prehistoric human dung.

Early hominids like Australopithecus who lived between two and five million years ago were half our size and had much bigger molars than we have. These humans probably didn't eat much meat apart from insects or reptiles, as they were not fast, agile or bright enough to catch much unless it was already dead. A couple of million years ago during the ice age Africa cooled down and fruit became scarcer. Our *Homo erectus* ancestors, in order to survive, now had to find better hunting and gathering techniques. Studies of chimps show they can take up to eleven hours to chew raw meat properly, so humans wanting better things to do with their time had to work around this. They initially developed stone tools to cut up the tubers, roots and raw meat into smaller pieces.

Then another even more crucial breakthrough occurred about a million years ago: using fire to cook food in a controlled way (our information comes from ash found in a cave in South Africa). This opened up many more possibilities, as cooked food reduced toxins and the incidence of food poisoning, and allowed much more energy to be extracted from food in a short time. Importantly, it freed up the valuable time we had previously spent collecting, then eating and digesting the tough roots we'd gathered and the occasional bit of raw meat.

Now that we were eating cooked food, we needed less of the digestive juices and enzymes as well as less fermentation time, so the lower part of our guts shrank accordingly. With the intestines using less energy and receiving more calories from the cooked vegetables and meat, our brains rapidly grew bigger and we became vastly more proficient at hunting meat, a great source of calories.

The few remaining hunter-gatherer tribes in existence allow us to explore their diets and their microbes as a possible window on the past and on our ancestors. Although the risk is that, once studied, they are no longer totally isolated. One such group are the Hadza tribe living at the site of early man's origins around the Rift Valley in Tanzania. They live in flexible and mobile groups of thirty to fifty and divide food duties by gender: the males hunt in small groups for game and occasionally honey, and the women gather plants and berries and dig for tubers. The hunting produce varies by season: it is very low in the wet season and increases in the dry periods when animals go to seek water. The tribe has little or no access to modern processed foods, medicines or antibiotics.

An eccentric colleague of mine, Jeff Leach, co-founder of the American Gut Project, lived with them for a good six months, following their diet and lifestyle to see how he and, importantly, his microbes got on with their diet. His only contact with the outside world was the satellite internet on his laptop that he used to write his weekly blogs.

He reported that while he at first maintained his Western diet his microbes had changed just slightly in the African environment. He then spent another few months eating exactly like the locals. He estimates that in a year they may eat over 500 different species of plants and animals. This included zebra meat, kudu, dik-dik, honey and many different roots and berries. He swabbed himself and everything else in sight, but was disappointed to discover that his microbes, although they had a better profile, were still 'Western' and not yet real Hadza.

Another group who closely resemble Stone Age hunter-gatherers are the Yanomami living in one of the remotest parts of the Amazon on the Brazilian–Venezuelan border. They still live essentially similar lifestyles to their ancestors' and are split into about two hundred villages of groups of around a hundred people, and some move camp every few years. They have no domesticated animals but live off a wide variety of crops comprising staples of cooked bananas and manioc (cassava), vegetables, fruit and insects, and episodically kill monkeys, peccaries, birds, frogs, caterpillars, grubs and fish. When sampled, they showed one of the lowest blood-lipid levels of any population and no signs of obesity.[18] Two independent teams managed to acquire the right contacts, permits and insect repellent and after parlaying with the chiefs of some of the remotest villages obtained stools from these rare populations. The results were both fascinating and worrying.[19][20]

The most striking result was that each tribesman and woman carried a vastly increased diversity of microbial species compared to Europeans. Each tribe also had its own extra 20 per cent of microbes that are totally unknown to us. Exploring just one bug in detail, E. coli (the most comprehensively studied bacteria), produced over fifty-six novel strains never seen before. Several common European microbes were completely absent in the tribes. For example, our friendly bifidobacteria that we get from yoghurt and that all Westerners have, was totally absent in all the Hadza and most of the Yanomami.

Both groups had an excess of Prevotella, which other populations whose diet is grain-based also have, plus many bacteria useful for breaking down plants. Oddly, in the West these same 'healthy' microbes have been associated with auto-immune diseases like arthritis. A microbial difference was noticed between men and women, which probably reflects their different roles in food collecting and eating. Men are in charge of the hunting and eat more meat at irregular intervals, and women can spend considerable time preparing the main staple food, manioc. These differences are not seen in Western populations, where women have the same access to supermarkets as men.

These two studies demonstrate that a group of microbes that may be unhealthy in one population can have the opposite influence in another very different environment. They also tell us that the microbial community as a whole is more important than one or two species. Moreover, they give us an idea of how many of our gut microbes may have become extinct since the introduction or farming, pesticides and antibiotics.

The sad fact is that we have a very depleted microbiome compared to our ancestors.

Meat, hearts and microbes

From the observational studies of meat eating so far performed in Western countries, there is no clear evidence that eating non-processed poultry meat is harmful, but eating red meat has been consistently associated with a small but increased risk of heart disease and cancer and an overall increase in the risk of dying. No proper randomised trials have been done – it's hard to force these diets on people over years – but reasonably good data now emerges from combining large observational studies.

Two large US cohorts of 84,000 nurses and 38,000 male health professionals were followed for a cumulative total of three million person years. They showed that eating just one extra red-meat serving per day increased globally the risk of death by 13 per cent for red meat and 20 per cent for processed meats, with slightly greater effects for heart disease and a 16 per cent increased risk of cancer.[21]

Shortly after, data from the European EPIC study engaged in following 450,000 people from ten countries indicated a modest 10 per cent increased risk of mortality with red meat. An even stronger risk of up to 40 per cent was seen for processed meats like sausages, ham and salami, or those unclassifiable meats in prepared meals.[22] Using this data, the Harvard group estimated that reducing everyone's meat consumption to half a serving or less per day (45g) could prevent 8 per cent of US deaths. In the UK, to produce similar benefits in men, this would mean halving their current meat intake.

Eating a bacon sandwich or a hot dog daily will reduce your life expectancy by two years or, more strikingly, by one hour per sandwich. The equivalent for a packet of cigarettes would be five hours. The caveat is that these results so far only hold for Europeans; a study of 300,000 Asians found that although red meat intakes are lower, they are increasing in parallel with increases in heart disease. But unlike the Western studies there was no obvious direct meat–heart-disease correlation.[23] Clearly, not all people suffer the same side effects of eating red meats, and there are other factors involved too.

With the collapse of the theory claiming that it was the fat in meat that caused the increased mortality, we need to look more closely now at other possible causes and at what our ancestors ate. We talked earlier of the work of the wandering dentist Weston Price who found that isolated tribes, and presumably our ancestors, often preferred the fattiest cuts of meat, which contain most of the nutrients and vitamins.

As we noted earlier, why some people in the West can eat large quantities of meat without problems and others develop heart disease and cancer has remained largely unexplained. One theory was that we all have different heart-disease-predisposing genes that interact with the meat in some way, but this has not been proven. In 2013 a series of experiments exploring our gut microbes totally changed our views on our relationship with meat.

Cardiologists have long suspected that a build-up of a harmless-looking but stinky substance called trimethylamine (TMA) – a porridge-like plaque in the arteries leading to heart failure, high blood pressure and heart attacks – is a major trigger of atherosclerosis. In fact, the harmful effects occur only when TMA is converted by the addition of an oxide molecule into its nasty sister compound TMAO. This TMAO is solid and odourless and sharks and some other fish have a lot of it. When fish goes off, the whiff is partly due to the solid TMAO changing back to TMA, which is smelly and liquid.

A US team from Cleveland confirmed this suspicion by measuring TMAO levels in the blood of several thousand patients. They found that those with higher than average levels had a nearly three-fold risk of major heart problems.[24] The team went on to feed rats a dietary source of TMAO derived from two components of red meat, choline and L-carnitine. They found that microbes in the rat guts were needed to convert the TMA into its dangerous form TMAO, which led to the atherosclerosis.

This was confirmed in humans when they fed omnivore volunteers an 8-ounce steak. The gut microbes greedily used the L-carnitine for energy, and in the process converted the TMA in L-carnitine to TMAO as a waste product a few hours later. What was fascinating was that when they repeated the experiment after giving a broad-spectrum antibiotic (which wipes out most types of bacteria in the guts), no toxic TMAO was produced. This showed definitively that specific gut microbes were feeding off the L-carnitine to produce the nasty amine. Potentially, this demonstrated that heart disease could be prevented by manipulating our gut microbes.

The effect of the antibiotic was temporary and a couple of weeks later, after continued meat eating, the TMAO was being produced again. Between individuals there was a big difference in how much TMAO, which can be measured in blood, was being produced. When the researchers looked at a group of vegans and vegetarians whose microbes rarely encounter L-carnitine or meat and fed them (apparently without duress) steak, very little happened and TMAO levels hardly changed.

This clearly shows once again why we don't all react in the same way to food. Vegetarians, as well as having different microbial profiles from meat eaters, also have some inherent genetic differences. The researchers found that people tend to have microbes that cluster into three or four communities, known as enterotypes, which can be thought of a bit like blood groups. Some of these increased risk and others were protective against the side effects of meat eating. The protective profiles included low levels of Prevotella and high levels of Bacteroidetes, although these are probably over-simplistic and still need to be confirmed in much larger populations.

Vegetarians are therefore protected, but if they were to change to a regular meat or L-carnitine diet for several weeks beforehand their rare meat-loving microbes would wake up, start reproducing and make TMAO in larger amounts.[25] These studies were done in vegetarian mice, but the principle is the same: eating or abstaining from regular meat can change your microbes for the better or for the worse. It should be possible, however, to improve the microbiome of a regular meat eater by having a meat/carnitine holiday. In other words, indulging in an occasional steak may not be harmful if you maintain a high-fibre diet. But should we be worried about L-carnitine rather than meat itself?

Fishy tales and bigorexics

Concern is one thing, outright banning another. One problem is that there is also L-carnitine in fish. Cod, sea bass, sardines, tiger prawns and squid, for example, have 5 to 6 mg of L-carnitine per 100 grams, though this is only a tenth of the amount in beef which has 95 mg. Fish eat plankton and they also feed off L-carnitine and produce TMAO. Everyone thinks that fish is good for you, and provides good vitamins like D and E. Most people know that the longest-lived people on the planet, the Japanese living in Okinawa, exist exclusively on a fish and carbohydrate diet.

So is fish necessary for a long life?

Parents get annoyed when their children reject it. My son only ever ate fish disguised in breadcrumbs, and we lovingly called it 'underwater chicken' until he learned that chickens are poor swimmers. Strangely, fish aversion is common in kids, who are at their fussiest between three and five years old, and as it has a genetic basis it can continue into adulthood, although it's hard to think of an evolutionary advantage as lobster allergy, for instance, is rare.

The main way fish gets into the health news is via stories of contamination by deadly pollutants like mercury, dioxin or polychlorinated biphenyls (BCP), which can affect babies or lead to brain damage or (in theory) cancer. Contamination generally affects longer-lived species like shark or swordfish and is not yet a major problem in smaller fish. I mentioned earlier how many lab studies find fish oils to have protective effects on the heart due to their major polyunsaturated omega-3 fats, and fish eating is widely promoted because of this. Nonetheless, the hard scientific data to back up the claim that fish is the perfect health food is, surprisingly, not that convincing.

There are no good trials of actual fish eating, only of supplements. Recent meta-analysis of all the trials of fish oils concluded that the benefits of fish-oil supplements are hard to detect and have been previously overestimated.[26] Meta-analyses of observational studies of fish eaters have shown a 17 per cent reduction in their mortality and a 36 per cent reduction in heart deaths, but these may well be biased by a general healthy-lifestyle effect.[27]

Other large prospective US studies have looked at people taking up fish eating in middle age and found a very modest 9 per cent reduction in heart deaths in women, but no benefit whatsoever to men.[28] This could be due to the crude observational nature of the studies or we may have overestimated the health advantages. In this case, the mild beneficial effects of the fish oils are perhaps being balanced by the bad effects of the L-carnitine and the microbes that feed off them.

So fish, while not being harmful and containing many good nutrients, may not be the secret of eternal life for everyone. You don't have to live in Okinawa to live for ever. Other exceptional long-lived groups around the world often eat little or no fish, such as the many Sardinian mountain villagers or the Californian Seventh Day Adventists.

The L-carnitine we ingest in fish and meat, as well as in milk and many other foods that can cause heart disease, is an unusual nutrient. It is produced inside most animals by a mix of two amino-acids, but can only be broken down and metabolised by microbes. It is widely promoted as a supplement on nutrition websites with the claim that it helps glucose metabolism in the body's energy cells, the mitochondria, to burn fat. Surprisingly, in short and unconvincing studies it has been tried in diabetics and (brave) heart-disease patients, with some reported success. It has even recently been marketed in the UK as a weight-loss drink to take before meals. The company that makes the product Full and Slim claims it acts like a gastric band.[29]

Carnitine is also much loved by fitness fanatics and bodybuilders, who some doctors refer to as 'bigorexics', meaning sufferers from a new form of eating and behavioural obsessive disorder. Users are encouraged to take 2–4 grams of carnitine a day as a fat-burning and muscle-building potion. In fact, this could be contributing to the increasing heart problems seen in this group, many of whom also abuse anabolic steroids. In a recent survey of gyms, 61 per cent of bodybuilders took L-carnitine supplements at very high doses.[30] To get 4 grams of carnitine per day naturally you'd have to eat a lot of steaks – over twenty, by my calculations. The average meat eater consumes around 120 mg of the substance a day, and yet vegans exist on only 10 mg a day and show no obvious deficiencies. Certainly, carnitine supplements seem both unnecessary and an extremely bad idea as far as your heart is concerned – and another example of how we take one isolated chemical nutrient out of context and promote it as super-healthy, while the opposite is true.

Going walkabout

If the paleo diet lacks much biological or evolutionary logic, that doesn't mean it is necessarily bad in the short term, particularly if eating more fruit and vegetables is encouraged at the expense of refined carbs. A real-life experiment in the 1980s took a group of humans back to Palaeolithic times and saw how they fared.[31] Australian Aborigines have suffered most from the loss of their hunter-gatherer lifestyle and have high rates of disease. Even today half of Aboriginal men in Australia will die by the age of forty-five, and there are no clear signs of improvement.

An intrepid Australian researcher, Kerin O'Dea, found ten overweight and diabetic middle-aged Aborigine volunteers living in a modern settlement and suffering a host of Western ailments. She persuaded them to return to the bush for seven weeks with her, to live off the land like their ancestors had done.

They went to a remote unpopulated area of Australia that had been their original tribal lands, near a place called Derby in northern West Australia. There they existed on a high-protein (65 per cent), low-fat (13 per cent) and low-carb (22 per cent) diet, living off three principal sources of food: kangaroos (which have little fat), freshwater fish, and yams for carbs. Other occasional treats came from turtles, birds, rodents, insects, other vegetables and honey. Despite the impressive grocery list, the unfit and out-of-practice hunter-gatherers only managed to take in around 1,200 calories a day. At the end of the ordeal they had lost an average of 8 kg in weight. Their blood sugars had normalised and their unhealthy lipids and triglycerides had dropped dramatically.

O'Dea admitted that the reasons for the (paleo) diet's success were hard to fathom, as it simultaneously reduced calories, increased protein, reduced carbs and increased exercise. This study was a one-off, but is mentioned on paleo websites as proof that the diet can cure obesity and diabetes. It is certainly impressive, but other strict calorie-restriction diets have produced similar results, albeit short term. Several years later, the fate of the Aborigines was unclear.

There are plenty of anecdotes about tough-minded individuals who have normalised their diabetes and reduced heart disease risk through severe calorie restriction and exercise, and sometimes by totally cutting out carbs.[32] However, although change is possible in some highly motivated individuals, in the long term the practical reality in most patients is very different. In one study, the very expensive nine years of monitoring and constant nutritional support of 5,000 US diabetics ended in dismal failure. The intensive-weight-loss group lost only 3 per cent more weight and failed to prevent any diabetes-related complications.[33]

The only clear success story comes from a massive social experiment in Cuba, where the population had to suffer a five-year economic crisis in the mid-1990s. Motorised transport ground to a halt and free bicycles were issued, food was limited and people ate local produce. As a consequence they exercised more and ate less but more healthily, and the average weight of the population dropped by 5.5 kg. This was followed by a major drop in the incidence of diabetes and a reduction of 53 per cent in heart disease.[34] Sadly, when the economic crisis was over they returned to their old ways and the health problems resumed.

The Aborigines exemplify the kind of population who, we believe, had been very healthy on their omnivorous diet and became very unwell when they changed to a Western diet. This suggests that eating protein was not the main problem. The other odd example is the Masai of East Africa who consume large amounts of meat and milk but few vegetables. A survey of four hundred tribesmen in the 1960s found little evidence of heart disease and they had low cholesterol levels similar to rural Chinese.[35] They must have had interesting microbes.

Groups of humans that have over centuries evolved to eat large amounts of animal protein and fat seem to experience few health problems, and this could be because their gut microbes have adapted with them. It is only when diets change dramatically in a short time and our core microbes haven't had time to evolve that problems develop.

I have learnt that for me an occasional piece of meat is probably good for my health, but I believe eating red meat every day or as part of a long-term restrictive high-protein diet is too much. As for white meat, if you disregard the fast or processed kind, no one has found anything particularly bad, at least healthwise, about the 50 billion chickens the world kills and eats annually. The chickens might disagree – they do of course suffer from regular salmonella and campylobacter infections due to their poor housing conditions and defective immune systems. Studying meat reinforced my decision to avoid processed foods that often contain meat of dubious quality or of uncertain origin. Having fish once or twice a week is probably healthy, and observational studies show pescatarians enjoy similar health benefits to non-meat eaters, so for the moment I'll carry on eating my seafood.

As the definitions of vegetarianism become vaguer and 'flexitarianism' is a bit too woolly a term too, all of us should now consider becoming reductionarians – if not for the sake of our health, then to reduce global warming. If everyone cut out meat just one day a week, we would all reap the benefits. Although we lack hard data to support this, switching to organically raised meat may also be a slightly healthier option, as you'd hope it would be hormone- and antibiotic-free and so the animals would have had healthier microbes as well as healthier lives. Importantly, although more expensive, organic meat is also potentially more carbon-neutral. But maybe, rather than frequenting gyms and eating steak, we should all have a go at being part-time hunter-gatherers. Our ancestors, having no year-round supply of meat, had to regularly rely on other sources of protein.

8

Protein: Non-animal

Beans, other legumes, seeds, nuts and mushrooms are perhaps the commonest sources of protein for non-meat and -fish eaters, and some of the other vegetables and grains provide a bit extra. Most vegetarians have no problem getting enough protein if they have a varied diet including these items. Beans and lentils are known around the world as poor man's meat, containing as they do all the key amino acids necessary to make protein inside the body. All of them are complex foods and protein is just one of the many ingredients that can affect our health and our microbes. Many vegetarians have turned to soy and its fermented product in the form of curds – known as tofu and sometimes called by its less organic name, 'textured vegetable protein' – as a different source of protein that can also come from grains. The demand for alternative-meat products has grown rapidly over the last few decades, especially in fast-food restaurants in the form of veggie-burgers. More recently sales of meat-free products have begun to drop off, perhaps because of health concerns and the resulting media attention.

Soy-hormone burgers and cancer

Soy protein was first produced from the soybean in the 1930s, and strange as it seems was used as a fire-fighting foam before its edible properties became apparent. Like other beans it is a mix of carbohydrates, fats, vitamins and protein. Although most beans have 20–25 per cent protein, soy is the champion at 36–40 per cent.

Hardened meat eaters dismiss soy and tofu as fake meat only eaten by the Japanese and vegetarians. What is not so well known is that Americans and Britons now consume nearly as much soy as the Japanese. This is not because of dramatic changes in our habits but because soybeans (or more commonly their protein extracts) are an additive used in two-thirds of processed foods. Many 'non-soy eaters' unwittingly get a detectable dose from milk and dairy products from ruminants fed on soy.[1] In the US, as for corn (maize), there are billion-dollar subsidies for soy farmers and the country is the world's largest producer of this genetically modified industrially produced crop. Most of the $4.5 billion US soy market actually goes to feed livestock which we then eat.

Soy is one of the most controversial topics in nutrition, with strong arguments on both sides for it being either the ultimate health food or a major cause for concern. Soy and tofu have been part of the natural Asian diet for centuries, and all soy products depend on a complex fermentation process that involves bacteria, fungi and yeast. There is now reasonable evidence of soy's mild protective effect against breast cancer and possibly also in reducing recurrences of the disease.[2] Similar but weaker findings are suggested for prostate cancer.[3]

Inconsistent evidence from observational and trial data suggests it also offers protective properties against dementia and Alzheimer's disease in Asians, but so far there's no similar evidence for Europeans. Soy in Asia is consumed differently: there it is more often eaten as the fermented product (such as natto, tempeh and miso), which alters its properties and could account for some of the health differences encountered. Some of the many longstanding health claims have now been disproved: for instance, it is no longer believed to be of any obvious use against menopausal symptoms or osteoporosis.[4]

As we have noted before in this book, the same foods can have very different effects in different people. Even when considering a single soy product, we have found that its effect in Europeans appears different from its effect in Asians. Soybeans contain unique antioxidant chemicals called isoflavones which are converted in the gut into active compounds known as endocrine disruptors (such as genistein); these mess up your hormone pathways and can modify your genes. This group of chemicals is thought to act like a type of mild oestrogen that could potentially increase cancer risk. Early in my career I enthusiastically explored this hypothesis, and I also published another observational study that linked global soybean consumption to national rates of pancreatic cancer.[5] This proved to be yet another demonstration of a false epidemiological association due to bias, and there is currently no good evidence of an adverse effect on the pancreas.

We now know that the isoflavones in soy don't affect oestrogen levels directly but do stimulate oestrogen receptors and have gene-modifying (epigenetic) effects. Thus they have the capacity to switch our genes on and off and subtly alter our hormone responses, possibly and worryingly modifying fertility, sperm counts and infant development. Given the very large doses many of us are unknowingly ingesting in processed foods or knowingly giving our babies in soy milk, we should be doing more serious studies on potential long-term side effects.

Gut microbes might be the key to producing the active soy compounds in our bodies as well as regulating how quickly they are eliminated. Asians, who have a different gut-microbe composition from Europeans, can break down the soy internally so as to produce greater amounts of its active isoflavone compounds.[6][7] In the US the soy food lobby (already heavily government-subsidised) managed, on the basis of rather shaky observational evidence, to get a health claim approved for soy protein being protective against heart disease. However, to get sufficient health benefits from soy you will either need to eat lots of processed or junk food or regular Japanese-style portions of miso soup, edamame or tempeh three times a day (roughly 100 grams in total).

Now, when it comes to modern processed foods, the problem is knowing what exactly they contain. If you are very lucky the label may give you some clues. Like other beans or legumes, soy is complex and has hundreds of ingredients. Some of these such as phytates are toxic and prevent the uptake of nutrients, while many others are potentially healthy (like fibre and unsaturated fats). But industrial processing often strips the soy down to its protein, which can then be reduced further to its component parts. Too much concentrated soy protein without the other natural parts of the bean could produce adverse effects; the truth is we really don't know. It could overwhelm your microbes, which expect to be interacting with the many complex components of the natural bean.

Soy milk sales have been increasing rapidly in many countries and are now the commonest soy product consumed. It is a good source of protein for children allergic to cow's milk, although the cure is rapidly becoming the problem: allergies to soy are also increasing, and soy alternatives are now available. The other components of soy such as endocrine disruptors like genistein, mentioned earlier, are regularly contained in infant feeds in amounts large enough to be a potential worry. Concern here is because the first three years of life are crucial for the normal development of a child, when the genes are constantly changing and fine-tuning their functions so as to produce new proteins.

Given that the isoflavones in soy have such significant and usually beneficial epigenetic effects in cancer, we should be more cautious when considering whether to give soy to susceptible babies. When you combine the epigenetic effects of soy with other known endocrine disruptor chemicals like bisphenol (BPA), which is in many babies' plastic bottles, you could be preparing a dangerous cocktail.[8]

Seaweed suppers

An unusual source of protein is seaweed, although to get health-giving amounts you would need to spend your day in sushi bars, since protein makes up only about 2 per cent of the content of marine algae; the rest is hard-to-digest carbohydrate starch. Seaweed comes in many flavours and colours, is an important source of iodine in preventing thyroid disease, and contains potentially beneficial antioxidants. It is another beautiful example of our human ability to adapt our digestion to accommodate new sources of food. This is a change that has happened in our recent history, brought to our attention thanks to the Japanese love of sushi.

Japanese who lived on or near the coast for centuries enjoyed eating seaweed in many different forms, for instance as additives to soups or salads or to wrap their raw fresh fish in. Like most Europeans, they originally lacked the enzymes to digest seaweed's complex carbohydrate starch. This meant it would pass straight through the gut without liberating any calories for the human or any nutrients for the microbes. Fortunately, the microbes in the guts of regular seaweed eaters gradually acquired the ability to digest it and obtain energy and nutrients from it.

The average Japanese has now made seaweed a regular part of their diet, and they consume an amazing 5 kg per person per year. This is nearly three times their consumption of dairy, which they digest poorly. Other Asian countries have also become hooked, and over two billion tonnes of seaweed are harvested each year for food. Edible seaweed is usually the brown variety such as kelp and wakame, and the red type such as nori is mainly used in sushi and as a massage gel and skin toner. Here is another example of the human capacity to adapt to different environments, once again demonstrating to us that our bodies may be programmed differently from others'.

Algae and GM humans

We may well wonder at the flexibility of our microbes in their capacity to digest plants. Just one single species of bacteria, Bacteroidetes thetaiotaomicron, contains an amazing arsenal of over 260 specific enzymes for breaking down different plant structures and over 200 related genes. In contrast, we human hosts have a puny store of less than thirty enzymes, showing how dependent we have become on our microbes.

Scientists have worked out one way whereby our microbes have managed to maintain this amazing level of diversity: by swapping genes.

Once upon a time there was a marine microbe (also from the Bacteroidetes family) called zobellia, that lived happily by feeding off marine red algae. One day he went on an adventure. From the fish he was riding on, he jumped straight into a human belly, which would become his new home. In the dark human colon he met other microbes, and in return for not eating him he kindly lent them some genes of his own which they lacked.

What this marine microbe was engaged in is known as horizontal gene transfer. The process of swapping genes is very common in bacteria, and accounts for their ability to resist antibiotics and fight off viruses. So microbes in seaweed-eating Japanese colons now acquired the ability to break down the seaweed, which benefited them as well as the human host.[9] We don't yet know how long it would take or how many buckets of seaweed for the average unexposed European to acquire the crucial marine microbes, although a few Welsh and Irish living on the coast may have them already.

We have recently discovered that humans contain at least 145 genes that have 'hopped over' in this way from other species, making us good examples of GM animals.[10] We may have inherited our blood group genes and some obesity genes from bacteria and algae.

The marine biologists investigating horizontal gene transfer looked for evidence of other aquatic microbe genes now residing in terrestrial gut microbes. Acting like Interpol, they tracked down the unique enzymes of marine species living happily and replicating in human guts in America, Mexico and Europe, sometimes far from the sea.[11] Their success proved that the process was not just a one-off, and confirmed the new-found ability of some of us not just to digest seaweed and other algae but to deal with any new food. More interestingly, this ability confers not only an energy and nutrient value, but offers health benefits too.

Seaweed contains a whole range of novel compounds, including proteins and chemicals like our friends the polyphenols, which act in helpful ways with anti-inflammatory, antioxidant and anti-cancer agents that our microbes can unlock. Part of the algae cell wall is important as a source of fibre, and can be broken down into helpful short-chain fatty acids such as propionate. A few small-scale trials in volunteers have suggested that seaweed can assist weight loss, offering the possibility that its fibre content reduces appetite. So eating seaweed may have kept the Japanese healthier and thinner than Europeans and resulted in less heart disease and cancer.[12]

As mentioned earlier, the Japanese and in particular the Okinawans in the south are the longest-living people on the planet and have the highest proportion of centenarians (743 per million). Their ingestion of seaweed could play a major part in this. The coasts of the UK and Ireland are not unlike Japan's in that we have over six hundred species of algae, which we know very little about. Unlike the Japanese, Britons have investigated only a few, but we do know that at least thirty are edible. Traditionally, coastal dwellers used to eat seaweed as a rich source of calcium and iodine. A seaweed that is still enjoyed today in Brittany and Wales is laverweed (*Porphyra*), which when mixed with oats is made into laverbread (in Welsh, *bara lawr*). In Ireland dillisk is still eaten as a snack, while carrageen (Irish moss) is used for jellies and puddings.

There is a new culinary movement called Gastrophysics that advocates using seaweed in cooking to replace flavours like umami instead of resorting to meat, salt or MSG for the same effect. In the UK and Ireland seaweed farming is attracting commercial interest, as a crop fertiliser but increasingly as food supplements. European and US production remains tiny compared to the industrial scale of seaweed farms in Japan, but demand is growing.

If you or your family are not regular fish eaters and don't live near the coast, you may not yet have the microbe genes and enzymes to allow you the full benefits of eating seaweed,[13] though it is likely that if you were to move to Japan or start eating enough sushi you would eventually acquire them. As always our microbes, which can produce a new generation every thirty minutes, react much faster to food than we can. The story of seaweed illustrates both the ability of the human body to change and the symbiotic relationship we have with our microbes.[14]

Magic mushrooms and fungi

Mushrooms are hard to classify: traditionally they were thought of as a vegetable although they are not plants, and in that they need to eat to survive they are probably closer to animals. They are part of the fungi kingdom, which also contains yeasts. Mushrooms actually consist of a collection of large microbes that usually settle on decaying matter which they live off to grow and reproduce. As well as also living on soil, plants and fruit, they sometimes live off us humans. They do this in our dark, moist places like on our feet and especially between our toes, as athlete's foot, or in our armpits or groins (jock rot). They contain no fat and generally have near-equal amounts of protein and carbohydrate. They are full of the healthy antioxidant selenium which mops up potentially toxic chemicals in our cells and they contain vitamin B, and if they have been sunbathing there is sometimes some vitamin D. They go well with meat or can act as a substitute for it as they stimulate the same umami taste receptors on our tongues that inform our brains that we are eating valuable protein.

Fungi can also live in our guts, as yeasts. We used to think they occurred only in relation to disease, but new sequencing detection methods have shown that in healthy people they make up about 4 per cent of the gut biomass – but about this kind of fungus we know virtually nothing. Fungi are only a problem when they run amok, unchecked by our normal microbes. Many fungi live happily and symbiotically with our microbes and us in our intestines. Some alternative medical practitioners are regularly and incorrectly diagnosing many vague symptoms as being due to an overgrowth of candida and offering bizarre and worthless treatments in an attempt to remove a natural part of our bodies. Our normal bacterial microbes are crucial in fighting off major fungal invasions. When under threat the microbes probably do this by signalling to our immune system. But when we take antibiotics or have immune-system problems we change the delicate balance, and consequently often develop fungal infections such as the yeast candida, commonly seen in the mouth and on the tongue.

Most women develop candida of the vagina (or thrush) at some point in their lives. This is normally kept in check by our bacteria friend lactobacillus, which is found in yoghurt. Thus yoghurt has become a popular internet remedy for candida. There have been few trials of its efficacy, but one Australian study recruited 270 women about to receive antibiotics and followed them for thrush, which a quarter of them later developed. They were randomised to take orally or vaginally lactobacillus probiotics or dummy products. Sadly, the probiotics did nothing for preventing the fungal infection.[15] Using full-fat yoghurt in the vagina hasn't yet been tested properly, but even if it doesn't work many women apparently find it soothing. Immunologists are currently working on genetically modified lactobacillus strains that will be able to counteract vaginal viruses and reduce infections like HIV/AIDS.[16] Whether vaginal yoghurt catches on as a health food remains to be seen, but it clearly has potential.

The Chinese have been using mushrooms as remedies for centuries. Although there have not yet been any human mushroom trials, studies have shown that feeding button mushrooms to mice for six weeks is beneficial. It increases their gut-microbial diversity and their Bacteroidetes species, and protects against gastric infections and inflammation.[17][18]

One fungus we commonly eat without usually knowing its origin is mycoprotein, more commonly known as Quorn™. This is grown in labs that began by taking the original fungus found in soil (*Fusarium venenatum*) and domesticating it in the same way that our ancestors did with many plants. Quorn has a high protein content, 44 per cent, and when blended with egg albumin mimics many meat products in texture. In Europe it is the commonest meat substitute.

Quorn had less success in the US, where soy is heavily sponsored, and it was severely criticised. It was marketed to the public as a mushroom-like product whereas it is in fact a completely different, unrelated, form of fungus. Despite the usual media scare stories that any new product faces, there has been no evidence of it doing any harm. Whether we take in the occasional bit of mould with our cheese or a tasty fungus, they are quite likely equally good for us.

In conclusion, for non-meat eaters nature provides a rich source of foods containing protein which, if varied enough, can supply us with nearly all the same nutrients that carnivores get except for vitamin B12. Our ability to digest them and produce the various necessary chemicals and hormones from them varies with our diverse resident gut microbes. Our adaptation for seaweed eating, thanks to our friends the microbes, is a great example of our combined flexibility and the cross-fertilisation of genes across species which makes us all 'genetically modified'. It is hard to believe, but drinking milk was once as foreign to us as eating seaweed.

9

Protein: Milk Products

In the UK in the 1970s Margaret (the milk snatcher) Thatcher became infamous for stopping free school milk for the over-sevens. This caused outrage and massive street protests at the time. Then the emotions dissipated, and milk became steadily less in vogue as its fat content was reduced and sales and school subsidies gradually declined.

Milk is a mix of many ingredients, a valuable source of protein (3 per cent) and calories and with a mostly saturated fat content of 2–3 per cent, plus many other nutrients such as calcium.

When I was young, cow's milk was considered vital for growing children and the most natural, healthy product possible. Fifty years ago at school my teachers made sure you drank it, even if it got a bit smelly on hot days. Milk was the first food to touch our lips, and for most of us our main source of food for the first year of life, so surely it had to be good for us? As people realised there were differences between human and cow's milk, and drinking cow's milk, after all, had started only six thousand years ago, attitudes changed. There were more and more reports of allergies and lactose intolerance, and confidence in cow's milk fell. The negative press about some of the fats in our diet only accelerated the decline: milk began to be replaced with alternatives like soy and, more recently, almond milk.

But are we right to be giving up cow's milk?

In the 1980s when the dairy industry was heavily supported by the government, an influential Chinese epidemiological study, discussed earlier in this book called the 'Cornell China Study' reported data from rural China, comparing data from each county collected a decade earlier with the current rates of over fifty diseases.[1] The team's leader Colin Campbell reported a strong and significant correlation with milk intake and high blood pressure that led him to conclude that we should all be avoiding the dairy products that vested interests were foisting upon us.

What wasn't clearly indicated was that sixty-two of the sixty-five Chinese counties they tested didn't eat dairy at all, and the three counties that did and also had high blood pressure rates were unusual. All the counties were in the north, near Mongolia and Kazakhstan, in areas very different in terms of climate, lifestyle and dietary habits, where drinking camel and horse milk is common. This illustrates again one of the many problems posed by the observational data produced by many similar studies that have been the mainstay of much of our previous knowledge about risk factors: namely, that they can suggest spurious associations. These associations could equally well have been caused by increased weight, high sodium intake or lack of vegetables, or even by the very different genetic make-up of the dairy-eating folk in the north.

The study gathered so much data that the researchers were able to produce thousands of correlations between different diseases and dietary ingredients. Given that by convention we accept an error rate of one in twenty, many of these correlations are likely to have been false by chance alone. The other frightening finding that Campbell used to make his point about the dangers of dairy was that the milk protein casein could in large doses cause liver cancer in lab animals. It was later pointed out that a similar experiment using non-animal protein had produced a similar result.[2][3] It is also clear that whatever these Chinese villagers were calling milk – often fermented yak's milk – it certainly wasn't the sanitised, pasteurised refrigerated sort we drink in the West. The findings also pose the wider question of why the other sixty-two Chinese counties drank no milk at all.

We are milk mutants

Since we have become farmers, the traditional story goes, we have not had sufficient time for our genes to modify our bodies to adapt to new foods. It is less than five hundred generations since farming began, compared to five thousand when we left Africa and a quarter of a million generations of selection and improvement since we left the chimps behind. So our modern post-farming history is only a tiny speck of evolution – surely not enough time to change?

This was until recently the prevailing view, when our understanding of the speed of change was influenced by our studies of global milk drinking. Only 35 per cent of the world can drink a half-pint of milk without feeling sick. This shoots up to over 90 per cent in northern Europe and down to 40 per cent in southern European countries. However, it is not the 65 per cent who have the problem, but the milk drinkers, who have successfully mutated and spread around the world probably in the space of just two hundred generations.[4] The earliest milk gene mutation so far discovered was found in the genes from the DNA of a 6,500-year-old body.

With the help of the microbe lactobacillus, babies produce the lactase enzyme that breaks down the lactose in their mothers' milk, but this stops working when they start eating solid foods. This means they can no longer digest lactose, which is a combination of glucose and galactose, sugars with a tough chemical bond between them. Lactose is pretty unique and well adapted for its purpose – it is virtually only found in animal milk, and is a great source of fats, sugars and proteins and other vitamins like vitamin D, and it contributes to brain development and calcium absorption (for bones). Early farmers around Turkey (or, some believe, Poland) were the first to discover that cow's milk, if made into cheese and yoghurt, could be transported more easily and consumed safely without causing nausea or vomiting.[5] The two foods were made by fermenting them with bacteria like lactobacilli which, as already noted, unlike their human hosts have the capacity to break down the lactose.

Suddenly these new mutants had a transportable source of food, protein and energy that gave them an advantage and encouraged them to increase their herds and travel further. This ready source of raw milk and a chance mutation in some farmers must have given them a survival edge, leading to the rapid spread of the gene through populations travelling north and west from the Middle East. It has been estimated that an increase of 18 per cent in fertility rates would have been needed to see these massive changes in our own genes. So the northern Europeans or their children, who lacked the lactase mutation, were less likely to survive to reproduce. No one is exactly sure why.

Milk may have increased survival rates by saving the lives of sick babies with stomach bugs, or may have helped in droughts or reduced infections from contaminated water. It may have improved fertility by shortening the gaps between babies by reducing the time needed for weaning. Whatever the reasons, the impact was dramatic and the genes spread widely. The major mutation happened in Europe, and other smaller ones occurred in Africa and the Middle East,[6] again showing us how we and our genes can adapt fairly rapidly in evolutionary terms to new important sources of food. The lactase gene mutation was a major event that allowed the gene to stay switched on permanently into adulthood.

More subtle epigenetic changes to our genes (the process whereby they can be effectively activated or de-activated in response to diet or environment) may also be significant.[7] These changes can occur within a few generations and can then become permanent. Although currently I can only speculate that they may have led to the lactase gene change, there is probably better evidence that microbes played a key role as well. Raw milk, by which we mean untreated and unpasteurised, is a rich source of nutrients that can support a wide variety of microbes, from lactobacilli and bifidos that confer health, to a few that can cause disease, and many others we don't yet know the function of.[8] Raw milk can be drunk on its own or used to make traditional unpasteurised cheeses.

Early milk drinkers may have had an advantage, not only from the obvious benefits from the protein and calories of the milk they ingested but from being able to expand their repertoire of personal microbiomes, which in turn improved their health and immunity. The lack of success of the gene mutation in the rest of the world is harder to explain, particularly in East Asia where less than 1 per cent of the Chinese have the gene. This may be related to factors such as the hotter climate, causing the raw milk to spoil before it could be drunk, which led to infections. This climate theory doesn't explain how the milk- and blood-drinking Masai managed to adapt, so this still remains a mystery

Another oddity is that 10 per cent of north Europeans failed to mutate and so lack the enzyme lactase and yet many can still drink a glass of milk without symptoms. Others have the lactase gene yet report being lactose-intolerant. How can we explain this? Lactose intolerance is now a common phenomenon; sufferers report bloating, stomach cramps, pain and diarrhoea on drinking milk or related products. Countries like the US with a mixed genetic pool from Europe, Asia and Africa have high rates of reported lactose intolerance, with supposedly up to 40 million sufferers according to some promotional websites. The truth is that it is impossible to diagnose easily or accurately, so the real rate is unknown.[9]

In the UK one in five people reports nausea and bloating on drinking milk, although fewer than a third of these have any confirmatory signs or tests. Remember that over 20 per cent of people report stomach upsets when taking placebos in drug trials. Most symptoms don't correlate well with the lack of the lactase gene or with the clinical test, which involves drinking 50 grams of lactose and measuring the effects in the breath or blood glucose. Many doctors dismiss most cases as having a psychological basis. Regardless of the gene status or lack of positive tests, we can say that many people have a perceived intolerance to lactose. These groups often unnecessarily avoid all milk products, more and more leading to calcium and vitamin D deficiency, particularly in children.[10]

While we have mainly focused on the problems of digesting the carbohydrate part of milk (the lactose), the protein component may also be to blame, especially for allergies. A slight genetic difference in the milk protein casein has been uncovered and exploited by one Australian company. Genetically modifying cows to produce different types of milk such as low-fat varieties is not new. Most cow's milk protein comes in the A1 casein form, but some cows can produce a slightly different variety that tastes the same, known as A2 casein. For people who have symptoms with the regular milk, A2 casein milk containing a different protein is now being produced from specially genetically selected cows. My guess is that although clinical trials are looking promising, this will just eventually transfer the problem elsewhere, but let's see.

Delhi belly and intolerance

Jenny and Mary are forty-year-old identical twins from Swindon in south-west England. They both drank milk as children without any problems and had milk in their tea and coffee as adults. Then when she was thirty-five, after an acrimonious divorce, on returning from a trip to India with a friend Mary started getting digestive problems. She had regular stomach pains that kept her up at night, and intermittent but severe bouts of diarrhoea. This carried on for several years; she went to see her GP, and eventually a psychologist. As part of our standard genetic studies that we perform on all our twins, we tested all the 500,000 or so variations in their 20,000 genes. As well as confirming they were identical, we found that Mary, like her sister, lacked the mutated lactase gene on chromosome 2 that most Europeans possess which would explain her current problems: she needed to avoid milk. When she did this, to her amazement after two weeks her problems stopped.

While a great result for Mary, what this didn't explain was why she hadn't had problems earlier, or why her twin sister with the exact same genetic make-up was unaffected, despite following a similar diet. Once again, their microbiomes may have been to blame. Studies of other patients with lactose intolerance have shown the same variable response to milk, but when patients with consistent problems have been given a normally indigestible prebiotic substance called GOS (galacto-oligosaccharide) which can alter your microbiome, they saw dramatic benefits compared to subjects given the placebo treatment. After two months their microbiomes had changed quite dramatically.[11]

When questioned more deeply, Mary remembered she had suffered a bad stomach infection at the end of her trip to Delhi and had needed several long courses of broad spectrum antibiotics. These powerful drugs must have altered and reduced the diversity of her gut microbes enough so that the remaining species couldn't cope with breaking down the lactose in her colon, leading to the common symptoms. So even small differences in the make-up of an individual's gut microbes will determine how they react to milk, even if they have the same genes. These changes may also affect how we respond to many other foods.

Does milk make you big and strong?

Most of us were told to drink our milk if we wanted to grow properly. If you look at a genetic map of the world and colour in all the countries with the lactase mutation gene for milk drinking, there is a clear correlation with height.[12] Southern Europeans, with generally low lactase gene levels, are shorter, although they contain some subpopulations with lactase gene mutations that are taller. But as we know, just finding an association doesn't mean that milk is the cause of better growth: it may just be a marker of something else, like wealth or general nutrition.

I am just under 5 feet 11 inches tall, my father was 5 feet 8, his father was 5 feet 5, and his father, born in 1876 in Russia, was just 5 feet 3, showing an increase of eight inches in four generations. This pattern, though anecdotal, is not uncommon. When in the past I visited mainland Europe I often felt like a giant next to older southern Europeans, while at the same time finding myself metrically challenged in Holland. According to our combined studies of over 50,000 European twins, height is over 80 per cent heritable (that is, 80 per cent of the variation between people is due to genes).[13] Later, from studying 250,000 people, we and over sixty other research groups discovered over 697 definite height genes, showing that thousands of genes of tiny effect (perhaps a quarter) contribute to our height.[14] So from this and from the traditional view of genetics you might think that there was little room for lifestyle factors like milk drinking to improve height further.

Yet when you look carefully at historical records you see the enormous variation over time in this 'most genetic of traits'. Height has dramatically changed at many other points in human history, possibly reaching a peak in the Middle Ages when many Europeans, like Charlemagne around AD 800, were allegedly six feet tall. Then, during the mini-ice age of the seventeenth century when we started moving into industrialised cities, we became much shorter again, the French *misérables* averaging just over five feet at the time of the French Revolution. We have slowly resumed growing, but some have done better than others.[15] The Dutch have become the tallest nation in the world. They have gained on average seven inches (18 cm) in just four generations, according to national statistics and surveys of military recruits.

But how can this be? For evolutionary changes to our genes to occur it should take hundreds of generations, unless a recent 'height mutation' occurred, as with the lactase gene. But this seems unlikely because, if so, we would have found it by now.

Just sixty years ago the Americans were the tallest nation in the world, but the average Dutchman is now just over 6 feet 1 inch, four inches taller than their US counterparts at just over 5 feet 9. The easy explanation is that the Dutch love milk and drink more than the Americans. When I visit the hospitals and universities in Holland, I find that most Dutch students still have a large glass of it at lunchtime.

When you look around the world at milk-drinking and non-milk-drinking populations based on the rates of lactase gene mutation, as we have seen there is a clear correlation with milk consumption and height. The Scandinavians and the Dutch come at the top of both league tables. Dutch milk consumption was around 6 million metric tonnes in 1962 and peaked at over 13.5 million in 1983; it has been slowly diminishing since then but is still at 11 million. And the average Dutchman or woman is still consuming twice as many dairy products per person as in the US. The average height of the population is now recognised by the World Health Organisation as a good general marker of the health and prosperity of the country.

In the early days of the colonies, healthy American army recruits were three inches taller on average and consumed about 20 per cent more calories than their (above average) equivalents in England, who were only 5 feet 4, and post-Revolution US recruits were even sturdier and could expect in 1800 to live about ten years longer. Until the 1960s the well-nourished Americans were the tallest nation in the world; then quite suddenly and inexplicably they stopped growing (at least upwards). In the 1950s the consumption of sugar and meat started to rise, until in the 1980s meat and vegetable intake fell, to be replaced by fats and sugars. US dairy production and consumption steadily increased during the 1950s and 60s and has tripled since 1970 – a remarkable feat. However, what isn't obvious from these total dairy figures is that milk consumption actually peaked in 1945 and has been falling steadily since, to around half of those peak levels, and this is particularly in school-age kids.[16]

In the US, could the dramatic shift in diet from drinking milk to eating processed cheese have had any impact on the nation's height and stopped Americans growing? Over the years, many US children have undoubtedly lost a ready source of healthy microbes and other nutrients, while gaining the calorie content from fat. Whilst the rest of the world grew, average, non-immigrant, heights in the US just stayed where they were. Small changes in calcium or vitamin D would probably not have been enough to halt growth, and US osteoporosis fracture rates had peaked in the 1970s.

In an earlier book, *Identically Different*, I speculated that the differences between US and European body frames could be due to different experiences in the generations that lived through the world wars. The first half of the twentieth century was a deeply traumatic time for many Europeans. Large migrations of people, wars, influenza, malnutrition, rationing and sometimes starvation were commonplace, such as in the Netherlands during the infamous Hunger Winter of 1944. When you look at the changes in height, you see a good correlation with the suffering of earlier generations, between 1900 and 1945. A stressful environment mixed with poor nutrition could have produced reversible (epigenetic) changes to the genes, sending a message to the developing foetus to grow quickly if it wanted to survive.

Could microbes, as well as genetics, be involved in this process? We noted earlier that raw milk has plenty of microbes, and that few people drink it nowadays because of the occasional risk of infection. This is where pasteurised milk came in. This process started to be used in the early twentieth century in a bid to eliminate diseases like brucellosis, listeriosis and tuberculosis, which could sometimes be transmitted in cow's milk. At the same time, rare but significant causes of acute food poisoning like E. coli were much reduced.

To kill harmful microbes, milk is briefly heated to 72 degrees C for fifteen seconds, then quickly cooled and bottled. Most people think fresh pasteurised milk is virtually sterile, but new genetic methods have shown us the reality of what is still alive. The heating process kills only heat-sensitive bacteria, and many are heat-resistant. Even the potentially dangerous heat-sensitive ones are not all killed – they just remain there in lower numbers. In fact, the microbial contents of raw and pasteurised milk are remarkably similar, and there are over twenty-four different families of bacteria including our friends lactobacillus, Prevotella and Bacteroides, as well as many others involved with our health.[17] So although the microbial quantities are much smaller, even pasteurised milk could still have effects on us via our guts. Although cats are not humans, it's interesting that a classic historical study in 1936 by Francis Pottenger showed that cats fed raw milk and meat could survive for much longer across several generations compared to those fed boiled milk. Here was a fine illustration of the benefits of living over dead microbes.[18]

Welsh rarebits

When I first met Tina and Tracey on a twin-research visit to London, I half believed them when they said they were non-identical. They were twenty-five-year-old blondes from Wales, but Tracey was four and a half inches shorter than her sister who was already a petite 5 feet 5. Their mother had told them – a mistake often made about twins – that as the midwife thought there were two placentas, they must be non-identical. Yet their faces and expressions were identical, and after finding out that when they were young their friends had often confused them, I had no doubt they were truly identical, not least because a third of identical twins have separate placentas. This was confirmed when their DNA showed a 100 per cent match.

The twins had always eaten the same types and quantities of food, and early on had been identical in height. They thought their subsequent difference had happened when Tracey was around eight years old and had suddenly developed a rare form of joint disease, juvenile rheumatoid arthritis. This caused painful swellings of the knees and wrists as well as episodic fevers. She remembered feeling rotten and paying frequent visits to the doctor, plus a constant feeling of tiredness. She was given anti-inflammatory drugs and slowly her joints improved, so that by the age of fourteen she had no pain or problems but found herself permanently shorter than her sister.

Clearly, there are many possible reasons for Tracey being so much shorter; for instance, she was on corticosteroids for a few months, which may have had an effect, if only temporary. The more likely cause was the long-term state of inflammation produced by the body which was aimed against the falsely imagined attackers in her joints, the 'auto-immune reaction'. This inflammation, according to my patients, is like having a permanent mild flu-like infection, and it is what accounted for Tracey's fatigue. Another side effect of inflammation and the heightened state of the body's defences is the disruption caused to gut microbes. Studies have shown clear differences in the microbiomes of patients in the early stages of rheumatoid arthritis that can't be explained by medications or changes in diet.[19] The main bug they found (Prevotella) seems to thrive in inflammatory conditions and drive out the other, normal occupants (such as Bacteroidetes).

Tracey's lack of growth could well have been due to changes in her gut microbes, as we know that germ-free mice don't grow normally, even when given extra food. The altered microbes could also have signalled to the immune system, which, thinking it was under attack, halted her growth.

So changes in our diets and the resulting changes to our microbes may have influenced the recent trends in height and made the Dutch much taller than the Americans. The increasing use of medications and antibiotics could also play a role, and we'll discuss these later. Human breast milk has been with us for millions of years and has generally had a good press, then a few thousand years ago many of us adapted to drink its close relative cow's milk. The PR for cow's milk, which has been hit by successive scares about fat content, allergies and intolerance, has not been so good. All the same, the evidence still suggests that for most people there's a slight health benefit to be had by consuming milk, yoghurt and cheese, while the story for butter is unclear. And, it seems, the cruder and less processed the product, the better.

10

Carbohydrates: of which Sugars

'Sugar is the most dangerous drug of the times and can still be easily acquired everywhere . . . Just like alcohol and tobacco, sugar is actually a drug. There is an important role for government. The use of sugar should be discouraged and users should be made aware of the dangers.' So wrote the head of Amsterdam's Health Service in 2013 at the peak of anti-sugar hysteria.[1] At the same time, several bestselling books were released, such as Robert Lustig's that described sugar as toxic.

This addictive poison is a particular carbohydrate combination of a fifty-fifty mixture of glucose and fructose – sucrose, in fact, and what we all commonly call 'sugar'. The fructose component, in particular, is the villain that is attracting all the attention. A persuasive group of doctors and journalists have assembled an impressive case for the prosecution, alleging that this sweet substance above all others is responsible for our current obesity and diabetes epidemic. Most of us are still confused by the fat and cholesterol debate. Should we now be worrying about sugar too?

Glucose, one of sugar's chemical components, is only slightly sweet and we don't eat or drink it on its own. Glucose is the natural fuel of our bodies: it travels around the blood supplying vital energy to our muscles, brains and organs, where it is used up by our cells to power all our processes and functions. The other part of sugar, fructose, provides the sweetness and is a natural component of all fruits.

About five years ago, as I mentioned, like many others in middle age I started paying more attention to my diet and general health. Based on what I had heard, I wanted to reduce my intake of saturated fats and my risk of heart disease, as well as lose a bit of weight. I started by changing my breakfast: no more coffee, toast, butter and marmalade with eggs and bacon on Sundays. My ritual was now a healthy helping of soy milk added to my combo of healthy low-fat, high-fibre muesli and All-Bran, washed down with a cup of black tea and a glass of natural fat-free, concentrate-free orange juice from Florida. I would also have low- or zero-fat fruit yoghurt a few times a week.

What could be healthier? After all, sugar is just empty calories and, weight for weight, has half the calories of fat. Should I really be worried? Well, probably yes.

Everyone now knows that a normal 330 ml can of Coke or Pepsi contains 140 calories and over eight spoonfuls of sugar and a Mars bar contains seven, and if you are crazy enough to eat a bag of toffee popcorn you will get over thirty spoons of the white stuff. Eating or drinking any of these treats may feel naughty, but you know what you are doing. I was carefully avoiding these obvious sugary villains, yet like many people I was being conned by the food industry. Food labels present sugar content in grams, so to convert it to something meaningful you divide the stated grams by four to get the equivalent in teaspoons, so 8 grams of sugar is equivalent to two teaspoons.

My 'healthy' low-fat breakfast cereal was giving me plenty of fibre from the oats, whole grains and nuts but was actually equivalent to eating 20 grams of sugar, or more obviously five teaspoonfuls of the stuff. On top of that, my small serving of low-fat non-dairy soy milk gave me an extra teaspoonful, and there were another 4 teaspoons in my small glass of expensive (labelled not from concentrate) 100 per cent pure Florida orange juice. In fact, it's not as pure as it sounds, as most of it is made by pasteurising the oranges, storing them in sterile vats without oxygen (or, by that time, much taste either) for months, then re-adding the taste as flavour packs. Incidentally, other, cheaper, orange juices made from frozen reconstituted concentrates are not very different in sugar content. They don't need any extra – there's plenty of natural sugar in the oranges. Anyway, that's ten teaspoons of sugar before I had included my twice-weekly fat-free yoghurt, which adds another five. But I couldn't taste anything like this much sugar because I was being fooled by the skilful chemistry, processing, texturing and added salt of the food manufacturers. I was also easily misled by the 'No added sugar' labels.

Everyone likes sugar. Even babies who have never been exposed to it are programmed to seek out the flavour, and the sweetness will uniquely soothe them if they cry or are in pain. The average Briton ingests around fifteen spoons of it daily, and many considerably more. It is a natural human response to seek out sweet non-poisonous fruits that we can use as a rapid energy and vitamin C source. Another probable reason why we have this reflex is so that we could gorge on fruit at harvest time, thereby stocking up on vital nutrients to last over the winter. Our ancestors would never have been able to even dream about having access to an unlimited supply of liquid fruit or honey all year round.

So is eating fifteen spoonfuls of sugar a day either natural or, more importantly, harmful for us?

In the old days, sugar was the stuff in a sugar bowl on the table that people used to add to their tea, coffee and desserts and it came in a rectangular paper bag. Those days are largely gone. We now actually add less sugar to our food because so much is already pre-added for our convenience. An estimated 65 to 75 per cent of all processed foods on sale now in the US and the UK have some added sugar.

Pure energy or pure con?

Fructose is the sweetest naturally occurring substance and considerably sweeter than glucose. It is present in nature only in fruit, but thanks to modern food wizardry is present everywhere. Until recently we were so concerned with fat being the bogeyman that sugar got an easy ride, under the clever marketing guise of it being loaded with energy. In processed foods, sugar was slowly but inexorably filling the gap left by fat.

Sugars, whether sucrose, glucose or fructose, have been called 'empty calories' because there is no additional nutritional value in them. The food industry has skilfully used the word 'empty' to imply that sugar is just pure energy rather than a major source of body fat. The marketing teams were conveniently forgetting the fructose component and focusing on the glucose half of sugar, reminding us that athletes all use high-energy glucose sports drinks. Confectionery like Marathon (now Snickers) and Mars – 'helps you work, rest and play'– were aimed, they said, at giving us the energy to get through the day, or even run a marathon.

Sugar could even fight illness. High-sugar-content drinks like Lucozade were heavily promoted (despite there being no proof) as aiding recovery from illness (and more recently from sports injuries), each bottle containing over 12 spoonfuls of therapeutic sugar. Breakfast cereals consisting nearly entirely of sugar (more sugar than puff, in fact) were also promoted as a great way for kids to kick-start the day. Apart from the minor irritation of causing tooth decay, it looked like cheap sugar was a great source of natural energy with no real downsides, if you were healthy, that is.

So my new 'healthy' breakfast got me off to a flying start with 10 to 15 spoonfuls of totally fat-free sugar, equivalent to two cans of Coke or Pepsi. My fibre intake was quite good, which might limit some of the damage by reducing the speed of absorption of the fructose and glucose; but the extra calories, empty or not, definitely were not helpful. I felt cheated.

Since the day I realised my errors, my trips to the shops have become educational. The 'Zero Per Cent Fat', 'High Fibre', 'No Added Sugar' or 'Five a Day' messages on products are often supersize and obscure the sugar content, which is hard to interpret without a magnifying-glass and a maths degree. The labelling around sugar is intentionally confusing – mixing amounts of carbs, natural and artificial sugars, agave syrup, corn syrup, fructose, and sugar from fruit (as if it were super-healthy) in different and obscure serving sizes and names and euphemisms.

Fruit juices alone now supply a daily average of 100 calories per person in Western countries, and most people believe these are healthy and an easy way to get your fruit portions and vitamin C. However, 98 per cent of them contain fruit juice concentrate with massive amounts of sugar, more than you would find in a Coke or Pepsi of the same size.

Twice as bad are pink and 'old-style' lemonades and ginger ales and many other fruit-flavoured drinks or fruit cocktails, to which even more extra sugar has been added. These contain up to ten spoonfuls per serving, and there are equally large amounts in foods such as organic yoghurts, where the manufacturers can cunningly call sugar 'organic fruit-based' or refer to 'organic invert sugar syrup'. Agave syrup is often added as an alternative and you may find that the packaging suggests it's healthier than sugar as it is made from tequila cactus (and pollinated by bats) and 15 per cent sweeter. However, sadly, despite its exotic origins, this is not an advantage because it is in fact its 70 per cent fructose content that gives it its 'magical' extra sweetness.

A stroll down the supermarket aisles tells you there's sugar in Hovis bread and other healthy-looking wholemeal brands. The average burger bun has so much sugar that it would be classified as a dessert if we didn't have a tiny gherkin on top to balance it. Each small serving of ketchup has a spoonful of sugar. There is sugar in steak pies, soups, tinned beans, lasagne, pasta sauces, sausages, smoked salmon and seafood sticks, healthy-looking salads, diet salads, muesli bars, bran cereals and ready-made curries. A bowl of canned tomato soup actually has more sugar (12 grams) than a bowl of Frosties cereal.

Basically, it's hard to find anything in a packet or tin that doesn't have serious amounts of sugar – again, if you can read the microscopic print on the label. If the sugar comes from fruit or another 'better' source, no matter, because if there is little or no fibre your body treats it just the same.

Why is there sugar in all these things? Partly, it's because we like it. Our tastes have apparently changed. We no longer want to add our own sugar from the bowl on the kitchen table. These days we want everything to be sweeter; we wouldn't like to go back to the good old days when fish tasted salty and dried fruits were a bit tart. As the food we eat becomes sweeter and more sugar is added to processed foods and juices, our thresholds increase and we need more sweetness to satisfy us. We also don't like the taste of low- or zero-fat foods, and because salt contents have slightly dropped recently those helpful food companies have added something to compensate our deprived taste-buds. Sugar.

Our liking for sweetness is partly cultural and partly genetic. Although all humans like sugar to some extent, there is considerable variation due to differences in the sweet-taste receptor genes that we discussed earlier. The propensity for obesity is strongly genetically related to our preference for sugar, and in 2015 we identified via huge international collaborations an ever-increasing list of nearly a hundred obesity genes, each with a tiny influence.[2] For some people just having the susceptibility genes is not a problem until they encounter certain types of food.

One study looked at over 30,000 Americans and at the variants they had in the top 32 obesity genes. People with the bad luck to have inherited over ten obesity-risk genes were particularly susceptible to the effects of sugary drinks. They effectively doubled their risk of becoming obese over the next five years if they drank just one sugary can each day.[3] We don't know yet why sugar seems to interact so strongly with our obesity genes but it does suggest that our bodies have a hard-wired need to seek out sugar, perhaps as a way of detecting edible carbohydrates. Interestingly, most of the genes interacting with sugar turned out to be those acting on the brain.

In our twin study with colleagues in Finland we found that nearly 50 per cent of the differences between sweet-tooth preferences was due to their genes, and the rest a mix of the dietary or sugar culture around them.[4] We also found a clear positive correlation between how pleasant people rated a 20 per cent sugar solution and how often they ate other sugary foods.[5] While genes in early life seemed to partly determine our sugar-related behaviour, subsequent exposure to high levels of the substance later in life also appears to have increased our sweetness thresholds and in the last few years to make us want yet more.

Individual governments have shied away from putting limits on sugar in food and drink, even more than they did for trans fats, preferring what they call 'voluntary discussions' with the food industry. When in 2002 the World Health Organisation first proposed to limit the percentage of total calories coming from sugar (for example, carbohydrates: of which sugar) declared on the label to 10 per cent, the industry reacted furiously. In the US the corn-sugar lobby petitioned Congress and managed to threaten the WHO with a withdrawal of funds. Not bowing to pressure, an updated 2014 WHO draft report has recommended that the original 10 per cent level proposed is still reasonable and that governments should aim at reducing this further to just 5 per cent, equivalent to a single can of cola.[6]

Without legislation these recommendations are of little use, as the average UK and US citizen is consuming over double this, and many teenagers much more. The industry has reacted predictably, attacking the evidence as weak and claiming that you can't lump all sugars together as unhealthy. In response to similar industry lobbying, the UK government, despite increasing pressure from its Chief Medical Officer, official reports from Public Health England and public concern, is also resisting any real change such as imposing limits or a sugar tax.[7] In contrast, in 2013 Denmark, having scrapped its previous saturated-fat tax, increased its small tax on sugary foods to levels that have started to reduce consumption. A 2015 US study estimated that implementing a tax on sweetened beverages could save $23 billion in health costs and earn $15 billion in revenue within ten years.[8]

The rapid and recent rise of sugar in the Western diet came about mainly for economic and political reasons. During the Cuban missile crisis of the early 1960s Cuban sugarcane supplies dried up and prices rose, and the US wanted to be self-sufficient. Burger-loving Richard Nixon believed that keeping food costs low was a government priority, to keep the public happy and stop the poor rioting. The government was prepared to subsidise cheap food and the food giants were happy to oblige. In response to a surfeit of cheap corn being converted to starch, combined with generous US government subsidies, high-fructose corn syrup (HFCS) started to be used on a large scale in the early 1970s. This is a mixture of slightly higher-fructose-content sugar (a 55/45 ratio of fructose to glucose) that tasted much the same as conventional sugar made from cane or beet. Because the US government wanted to protect the corn business at all costs, they added extra taxes to imported sugar and ensured that US corn-made sugar was cheaper. This meant that it could be added cheaply to soft drinks and processed foods, thereby improving sales at little cost.

In Europe, the EU didn't want to use corn-sugar as it subsidises the local (mainly French) sugar beet industry (beet is a root vegetable). It does this in two ways: both via its infamous Common Agricultural Policy, so as to keep beet prices for the farmers stable and costing taxpayers over 1.5 billion euros a year, and by putting a 300 euro per tonne tax on any imported sugar cane, thus doubling its cost. The UK, thanks to its old empire, used to be able to rely on cane sugar, and even a company as synonymous with sugar as Tate & Lyle has sold its cane sugar business, mainly thanks to EU policy. The net result is that sugar across the world is cheap and, ironically, heavily subsidised by taxpayers. This has contributed to the boom over the last thirty years in sales of sugar-enhanced drinks. Liquid calories for the first time in our history now make up a major part of the Western diet.

The question over whether it's sugar or fat that is the cause of our modern obesity problems was debated fiercely in the 1970s. A British physiologist and nutritionist, John Yudkin, who was the most vocal critic of the Ancel Keys theory of high-fat diets causing disease, in 1972 wrote a prescient book called *Pure, White and Deadly* in which he presented his case that sugar, not fat, was the main enemy.[9] At the time neither Keys or Yudkin, who became sworn adversaries, had access to decent clinical trials, and both relied on potentially flawed observational epidemiological studies for their arguments. In the end Keys was a much better politician than Yudkin and 'won' the debate, at least as far as governments were concerned. Any sugar concerns were pushed under the carpet for the sake of a clear anti-fat message that suited the major players.

Yudkin's point was that refined sugar was a relatively new addition to our diets and that unlike in the case of fat we were now eating twenty times more sugar than ever in our history. Before the advent of early farming we could get sugar only from eating ripe fruit or wild honey, so for most people this was a rare event. As agriculture developed, sugar cane started to be grown, but it remained costly to produce and like honey was a luxury item. In the sixteenth century it was equivalent to buying caviar today. Plantations in the Caribbean, helped by the slave trade, started to increase the production and quality of sugar cane and slowly and progressively brought prices down.

The recent changes to sugar consumption per head are hard to estimate accurately because of so much now being added to food, but it has increased around twentyfold since the end of the nineteenth century. Since 1990, total sugar consumption in Britain has increased by 10 per cent per decade, mainly at the expense of total fat.

The question remains: Are these empty calories which, gram for gram, we eat more of than fat or protein, better or worse for us?

Busy tooth fairies and deadly mouthwash

'I thought giving him juice in a bottle was good for him.' Billy's parents believed fruit drinks full of nutrients and vitamin C would be healthy, but they inadvertently triggered a spiral of decay in their infant. Despite brushing twice a day with fluoride toothpaste, the five-year-old's teeth have completely rotted. Billy had ten teeth removed under general anaesthetic at Manchester Dental Hospital, the maximum allowed under current guidelines. Surgeons wanted to take out fifteen, but that would have meant nine months on a waiting list and an overnight hospital stay. Billy now has just ten teeth left – four at the top, six at the bottom. The rest of his milk teeth, which were decaying before they even came through the gums, will fall out in the next six months to make way for his adult teeth. Meanwhile, he is no longer allowed to take fruit juice to bed.

Billy first had problems when he was two and a half. His mother, a twenty-five-year-old computer engineer, said: 'I had never received any advice on how to look after Billy's teeth. I know it's probably my fault but my actions were innocent. I think Billy has a balanced diet and although he doesn't drink that many juices or fizzy drinks or eat that much chocolate, I'm told now it's the time of the day he had them that was bad. Bedtime was the worst, so the sugar had all night to work on his teeth.'

Five hundred kids in the UK need hospital admission every week to remove rotten teeth. Tooth decay is affecting over one in ten five-year-olds, and the current juicing craze is increasingly causing problems in adults. The first group of professionals to complain about sugar's side effects were dental researchers, who after a dip during the Second World War noticed a surge in dental caries (the polite Latin word for 'rot') when sugar rationing ended.[10] Mothers were still adding sugar to formula feeds and to dummies and soothers, but many dentists were happy with the extra work and rewards that filling cavities brought. They didn't try too hard to alter their patients' dietary habits but would sometimes tell them off for not brushing enough. Dentists' families, in contrast, didn't have sugary drinks at night, not even milky ones, and seemed magically never to suffer the same problems, showing that the decay was entirely preventable.

The epidemic that peaked in children in the 1960s dramatically ended as decay rates plunged 5 per cent a year, when countries introduced fluoride in water and toothpaste to combat the sugar. Some dental researchers like Aubrey Sheiham, whose lively lectures I remembered at medical school, were vocal over the lack of action by both the profession and the government against the increasing sugar consumption. By the mid-1980s, because of sugar, tooth decay in developing countries had now overtaken the West, where rates had halved, so Sheiham suggested our dentists should either 'change countries or start playing more golf'.[11] So by the 1960s we already had clear warning that not only were our bodies not adapted to this novel diet, but it was causing us harm.

In the West, while tooth decay was declining dramatically, some dentists were noticing the same in regions without fluoride or major changes in tooth brushing, and thought this could be due to the increasing use of antibiotics in children.[12] It transpired that cavities were actually caused not directly by the extra sugar, but by microbes.

Our normal microbes were just not used to the abundance of sugar, but one particular species, called *Strep mutans*, loved the new food and fed hungrily off the sugar around our teeth and gums and quickly multiplied. Unfortunately, unlike the other harmless microbes, they used the sugar to produce lactic acid which made little holes in our tooth enamel. The *Strep mutans* stuck to our teeth by attaching themselves to dental plaque. This substance so familiar to us is in fact a colony of six hundred species of harmless bacteria stuck together to form a sticky mucous commune (called a biofilm). Ingeniously they produce a glue-like substance from the sugar they metabolise, which allows them to keep feeding safely while they hang on. Ironically, people who use a daily mouthwash may be killing off their healthy microbes and allowing the harmful ones to take over, leading to even more gum and tooth disease.[13] One small study suggested this habit also increases blood pressure and risk of heart disease.[14]

Even at the height of the cavity epidemic, around 15 to 20 per cent of children would be relatively untouched. They are strangely protected, even if they have sugary cereal and a cola for breakfast and hardly ever brush their teeth. This is because they are lucky enough to have genes that make special saliva proteins which inhibit *Strep mutans*'s eating habits.[15] My brother and I lacked these genes (by which I mean the good variants), and as young kids experienced the tooth decay epidemic at first hand. We regularly competed with each other to eat the most bowls of breakfast cereals we could manage so as to get a brief sugar rush and feed our microbes before starting to feel very sick indeed.

You may remember eating Sugar Smacks, Honey Smacks, Sugar Puffs, Coco-Pops, All Stars, Frosties etc., but what you probably didn't realise was that they contained over 35 per cent pure sugar. In the US the same brands contained an additional 10 per cent sugar. We spent many happy hours at our local dentist, who was quite content for us to keep eating the cereal while building a large swimming pool at our expense. What is worrying is that forty years later these same cereals (some brands have dropped sugar from the name) are still being sold with healthy-looking nutrient packaging and no warnings. After decades without any tooth problems other than those caused by my over-zealous Australian dentist, I now have a couple of small cavities. This could be just bad luck, or may be caused by my extra sugar intake from my super-healthy low-fat breakfasts?

Recent research shows that probiotics, the friendly bacteria like lactobacillus, offer some protective effects against the acid-producing *Strep mutans*. A German company has developed a probiotic (sugar-free) sweet which when sucked five times a day reduces the numbers of the microbe.[16] The probiotic is a lactobacillus similar to those found in cheese, but preheated so as to kill it. It still binds onto the mouth microbes, cramping their style, and prevents them sticking to the plaque around our teeth. They then get flushed away by the saliva. Longer trials of similar probiotics show they continue to survive in the mouth, providing benefit, for several weeks.[17] Trials show that giving kids more natural microbial cheese or sugar-free yoghurt could do the same.[18]

Tooth decay, despite fluoride, is making a comeback. Rates are again increasing in most countries and about a third of the planet has untreated tooth decay.[19] [20] The only people without problems are rare tribes who live off meat or fish, as our hunter-gatherer ancestors did. Even our early Neolithic farming ancestors on their starch diets we now know suffered from tooth decay. In the UK dental extraction is now the commonest cause of admission for kids, and costs the country over £45 million. Increased sugar in the form of fizzy drinks and juices is the main culprit, and now together with our tooth microbes is overwhelming the protective effects of fluoride.

We know a lot about the effects of sugar on the bacteria in the mouth, but much less about the effects on the gut. This is because most studies have either focused on high-fat diets or on mixtures of high fat and high sugar. Given that our ancestors only rarely had access to honey and didn't do smoothies, our system and our gut microbes are less well adapted to high doses of sugar, particularly as a liquid.

Chewing or sipping your carbs?

Our whole digestive system is set up as a planned sequence of events that trigger, then manage, the digestive process. It starts with the brain just thinking about food so as to get the gastric juices and the hormones flowing, as well as producing the amylase enzyme in saliva. Then comes chewing. The body expects us to chew and swallow our food slowly. Up to forty chews is believed to be optimal to break down tough meats and vegetables and to prepare the whole digestive system. Compared to our ancestors, nowadays we rarely use our chewing powers and jaw muscles to the full, as shown by the lack of growth of our jaws and the subsequent modern epidemic of impacted wisdom teeth resulting from a mismatch of jaw sizes.

Normally the well-chewed and broken-down food passes downwards, releasing hormones in the gut lining, the liver, the pancreas and the gall bladder that aided its breakdown. At the same time signals of fullness are sent back to the brain. The pancreas releases insulin so that any glucose released into the bloodstream can be quickly dealt with. The gall bladder releases bile salts which in turn send signals to the microbes further down the gut, in the colon, to get ready to digest the expected food.

So when you drink a large sugary drink with, say, a bowl of

refined-carbohydrate pasta that requires minimal chewing, your body has no time to send the right signals. As the high-sugar load hits the stomach it passes rapidly into the small intestine, where most of the sugar is absorbed. This produces an abnormal and mistimed insulin response, altering the breakdown of glucose; the released bile salts now have the wrong mix for the unexpected sugars, and the normal gut microbes are replaced by unhealthy species that feed off the sugar scraps. These unusual microbes send out new messages to alter the hormonal signals and the bile salts. The result is a very disturbed system. The microbes expecting nutrients from the empty-calorie food send out signals to the brain to keep sending more sugar their way, while the glucose is being stored away as fat, often of the internal (visceral) kind.

Can our microbes defend us against fructose?

The reason for fructose being the latest hate item of the diet world despite its natural 'found in fruit' credentials is complex. Forty years ago Yudkin pointed out in his book that fructose was the likely villain, and that the glucose released from starch in plants was very different. He was largely ignored. However, the high-sugar content of soft drinks did eventually grab people's attention.

George Bray, a distinguished obesity researcher, reignited the sugar debate in 2004 when he pointed out the clear observational correlation between rising sugar ingestion and rising obesity in the US.[21] In most countries, the consumption of sugary drinks had increased three- to fivefold since 1950, and by 2009 around 20 per cent of total calorie intake in the UK came from high-fructose soft drinks (more than that, in some teenagers).[22] Across the world, these changes mirrored increases in the incidence of obesity and diabetes.[23] Confirmatory epidemiological evidence from the meta-analysis of other large observational studies showed that the consumption of soft drinks was associated with a later risk of obesity and diabetes.[24]

When fructose is compared to glucose it shows a number of major and worrying differences in how it is metabolised. Most of it is absorbed in the gut and goes straight to the liver, where it is converted to glucose, energy or fat. But unlike glucose it produces only a very modest insulin signal in the blood. At first, cocky doctors recommended fructose sweets for diabetics, which was a really dumb idea. Fructose does work differently, but it interferes with normal appetite signals to the brain. We know little about how fructose and glucose remnants in the gut interact with our gut microbes, but many sports drinks contain fructose and this triggers in a growing number of people microbial fermentation, bloating and discomfort.[25] This fructose intolerance has a genetic basis; these individuals can't process the fructose, which builds up to high levels in the blood, something that is hard to achieve in nature.

The important differences between fructose and glucose metabolism triggered a number of rodent studies that were tricky to do in humans. Fructose in rats causes the same toxic change in microbes that we saw with junk foods and high-fat diets – in particular, causing a fatty liver. But this side effect can actually be reversed by antibiotics.[26] When rodents are fed high doses of fructose, it increases their visceral fat dramatically.[27] Randomised trials of fructose in humans have been less clear-cut, but often show metabolic and visceral-fat changes after a few months.[28] This internal fat effect of fructose and soft drinks is likely underestimated, and could be responsible for the diabetes epidemic in places like the Middle East where huge numbers of people consume soft drinks, often without looking externally fat. These people are called TOFIs (thin-outside, fat inside) and are metabolically very unwell.

Fruit contains high levels of fructose, so should we be worried about overeating fruits? There is a lack of good data on this subject, but it appears that eating the whole fruit is much less harmful. A study of 425 Japanese living in Brazil at high risk of diabetes found that the fruit eaters had normal insulin increases, in contrast to the doubling of insulin peaks in those that had acquired the same amount of fructose by consuming sugary drinks.[29] Other smaller, detailed, studies showed similar results, and that there is something else in whole fruit that is protective. This is probably the increased fibre, which we discuss later.

Eating or drinking too much sugar in any form is harmful, and liquid calories appear particularly bad, even if dressed up as 'healthy juice'. However, despite the hype, the evidence that fructose is the principal demon to be exorcised is not proven. Its metabolism and the way the body handles it is certainly different from glucose, and although theoretically it could be much worse for us we should remember that glucose overdose also leads to fat deposition.[30]

The critics of the fructose witch-hunt argue that the data is flawed, and for several reasons. These include, if total calories are controlled for, that fructose is no worse than other sugar sources; that the rodents studied are fed abnormally high fructose diets (60 per cent of calories); that rodent livers are not the same as humans'; and that the human studies are generally small, of poor quality and inconsistent.[31] This renders inconsistent even the academics' interpretation of the same trials by meta-analyses.[32][33][34]

Healthy lean people seem to be able to deal with occasional fructose drinks with no problems. Because humans in most cases are not rats, not enough money has been spent on careful trials. As a consequence, we still can't prove that the excess calories from fructose are actually worse than the excess calories from glucose. Until we know more, then, we need to avoid the same mistakes of reductionism and avoid, too, singling out scapegoats like fructose, which takes our eyes and minds off the bigger picture. While I am now convinced that too much sugar is not good for us, particularly if it comes in liquid or unnatural form, the data is not yet clear that fructose is much worse than the equivalent amount of glucose or other sugars.

Looking back on my past ideal breakfast, it would seem that my high-sugar feast of muesli, fruit yoghurt and fruit juice should be replaced with a strong black coffee and some natural yoghurt. In fact, I'm beginning to wonder if I should go all the way back to my porridge, eggs and bacon.

11

Carbohydrates: Non-sugars

Fergus was a farmer living near Cork in southern Ireland. He had retired late at the age of seventy and now had no more responsibilities. Being fit and healthy, he was looking forward to relaxing in his idyllic surroundings with Mary, his wife of forty-eight years. One day she noticed a lump on her breast and was diagnosed with metastatic cancer. Six months later she died. He became a bit of a recluse, kept close to his cottage, rarely went to the local town and didn't mix with his neighbours. One sunny day three years later the local GP, with rare time on his hands, decided to visit him. Fergus had been on the practice list for years but had only ever visited in order to support his wife; he had never made a personal appearance or, apparently, needed to see a doctor.

The GP had remembered a very healthy man for his age, who was slim, didn't smoke and was physically active. He was shocked at the change in him. His skin was sallow and grey, he had lost some teeth and he looked fragile. 'I think you need to come in to the surgery for a check-up.'

Fergus now had high cholesterol, mild diabetes, high blood pressure and arthritis of the hip that gave him a limp. Also, his memory seemed to be fading.

The GP puzzled over the reasons, and asked the two most common questions doctors ask in Ireland: 'Are you depressed?' and 'Are you drinking?'

'Well Doctor, you are right,' Fergus replied. 'I was real upset and tearful and took to overdoing the drink. But only for the first six months. Then I pulled myself together, gave up the booze except for the occasional Guinness, and I'm OK now.'

The mystery of his recent deterioration was finally unearthed by the practice nurse. His wife had previously done all the cooking; he couldn't boil an egg and was too proud to ask for help. He had lived on tea and cheese sandwiches for the last three years. Fergus moved to a local nursing home a few months later. Although his diabetes was treated, he died in his sleep of a heart attack within six months.

A microbe collaborator of mine, Paul O'Toole, who was working in Cork at the time, told me this story as typical of many in the area, in which a dramatic change in nutrition in an elderly person had preceded health problems. Paul's research group is looking into the effects of the microbiome in the elderly, and particularly the effects of diet. In an important study they surveyed 178 Irish residents from local nursing homes aged 70 to 102, half of whom came temporarily for day care and the rest who lived there permanently.[1]

They found that within six months all the permanent residents on the same dull, institutional diet developed a similar microbiome. Unfortunately, it wasn't a healthy mix and was lacking in diversity and many of the healthy microbes. It also indicated greater levels of inflammation. Those elderly part-time residents who sometimes cooked for themselves and ate non-institutional meals had healthier microbiomes than those with the communally prepared food. There was some variation, but within a year of entering the long-stay nursing home all the residents had very similar and unhealthy microbiomes.

In elderly people there are many complex reasons why the body deteriorates. These include loss of muscle through lack of physical exercise, depression, social circumstances and loss of cognitive function. The loss of teeth, changes in saliva, and the increasing use of antibiotics and other drugs can also affect the microbiome. With age, we see increases in the number and dysfunction of the protective Treg cells, which as we know interact with our microbes and at this point in life can over-suppress the immune system. Even when all these other factors were accounted for, diet and nutrition remained the dominant factor in determining the microbiome and its relation to health in the elderly. The residents with the least diverse gut microbiomes were far more likely to be frail and suffer from illnesses – and whatever the cause, were more likely to die within a year.

When Claire Steves in my lab looked at 400 of our older UK

twins living independently who were generally frail, she found that they had a less diverse microbiome than average and lower numbers of microbes associated with suppressing inflammation and gut leakiness, such as *F. prausnitzii*. They also had fewer beneficial lactobacilli. This was the same result as an earlier, smaller, study of frailer old people had produced, so unlikely to be just down to chance.[2] We think it likely that the changes in diet led first to the changes in microbes and then the frailty, rather than the other way around.

The exact deficiency in the residential diet that had had such a dramatic effect was unclear. Alcohol drinking at the residence is not encouraged. Although drinking the barley-rich Guinness was once thought to be healthy, the company has not used the health claim for many years, so Guinness deficiency is unlikely to be the major factor. Apart from the obvious monotony of the diet, the main culprit appears to be a lack of regular fresh fruit and vegetables – carbohydrates, in other words.

Carbohydrates come from plants and fruit, and in many different forms; they vary in how easy it is to extract energy from them. Confusingly, they are also called 'saccharides' after the Greek word for sugar. When the molecules are small, that is, one or two joined together, they are known as monosaccharides and disaccharides, respectively; they are what we commonly call sugars and are used in processed foods. When longer they are called polysaccharides and are used either for energy storage or for maintaining the structure of the plant, its fibre.

Most of the edible carbs in our diets are called starches, which are the main energy store for plants and the main ingredient of potatoes, bread and rice. They consist of groups of glucose molecules in long, tightly bound chains. Some forms are easy for us to break down and others are more resistant. As we saw earlier, humans have just thirty enzymes to break down all the complex carbohydrates they ingest, but luckily our gut microbes have over six thousand at their disposal to do the job properly. Without our microbes waiting to get to work on them, most of our carbohydrates would be simply a good workout for our jaws.

Raw-foodies and the toxic tomato

Paleo food enthusiasts claim – falsely, as we discussed previously – that we haven't evolved sufficiently to eat the new foods produced in the last ten thousand years, and raw-food advocates go further to apply a similar logic to avoiding cooked foods, which we have eaten for up to a million years. The raw-food movement promotes a form of veganism that claims that cooking foods removes much of their natural nutritional value and destroys useful enzymes. The movement comes in many forms and in different degrees of strictness and orthodoxy, including fruitarians, juicearians and sproutarians.

Some of the more liberal of these groups allow vegetables to be slowly warmed up to 45 degrees C, which doesn't destroy any of the nutrients or enzymes. This involves lightly cooking the food for long periods, and there is some scientific basis for the method. A few high-end restaurants have started to use *sous vide* methods, involving slow cooking at around 55 degrees C in a vacuum. My experience of this recently in a Brussels Michelin starred (and inevitably pricey) restaurant was of amazing flavour and texture. I visited the kitchens of Comme Chez-Soi afterwards. Having no grills or ovens, it looked more like a space-age lab than a kitchen.

Anyway, most raw-foodies usually don't eat meat and are more interested in the health benefits than the taste. According to them, the 'valuable and intact' enzymes are precious to us. This is despite the annoying fact that our digestive processes effectively and promptly deactivate them. Clearly, we have continued to evolve during the last million years and those diehards who kept on eating raw food died out long ago – and for good reason – it's hard to extract sufficient calories and nutrients without cooking it at least a little. With the development of cooking we gradually lost over a third of our intestines, along with the ability to exist on raw food. In the modern world this could be a good way to slim down, not because of the benefits bestowed by the magic enzyme ingredients but just because we can't efficiently break down the complex carbohydrates into energy.

Restrictive raw-food or paleo diets clearly have some advantages, and reducing refined carbs and avoiding processed foods is a big plus. However, reducing choice and diversity by excluding whole ranges of foods is a big mistake. For example, as mentioned already, paleo diets point the finger at poor old tomatoes as they come from a large family of mainly toxic nightshade plants to which 'we haven't had time to adapt'. This is daft. Saying the tomato or its family is the cause of auto-immune disease is also very misguided, and using pseudo-science to focus on side effects of one or two of its hundreds of chemicals is worse. No studies have convincingly implicated tomatoes as the cause of disease, and they are an essential part of the Mediterranean diet, the only one proved to reduce heart disease. Moreover, other more rigorous studies show that one of its many chemicals, lycopene, may help to prevent cancer, which is certainly not a good reason to eliminate it.[3]

While a lack of diversity and particularly of fresh produce and their nutrients is bad for us, at the other extreme the Web is bursting with colourful stories of people who exist on nothing but raw fruit and nuts – the fruitarians. They claim miraculous powers and energy, yet only very few can sustain this without spending many hours in the kitchen chopping and juicing, or on the toilet.

One exception is Freelee the Banana Girl, a former bulimic from Adelaide who for the last decade has been eating 90 per cent of her diet as fruit, with the odd cooked vegetable thrown in. She is now an ultra-toned fifty kilos, and according to her website lives in Queensland in a skimpy bikini surrounded by bananas and photographers. She has a popular video showing her eating fifty-one bananas in one day, washed down with coconut milk, which in theory gives her an intake of over 4,000 calories.

Yet despite not limiting her intake of calories she stays slim. She does admit to usually only eating twenty-odd bananas a day, but will have more or other fruits if she feels peckish. She always stays on raw food till 4 p.m., then may have some lightly cooked vegetables for dinner. She has concocted a range of diets and cookbooks and the thirty-bananas-a-day diet is widely promoted, with reports of both spectacular successes and failures.

Before you sign up for her health plan you should know that she also believes that losing your periods from dieting for nine months is good for you, and that fruit not chemotherapy is the treatment for cancer.[4] Other fruitarian advocates were the late Steve Jobs, whose company was clearly influenced by his diet, Mahatma Gandhi and reputedly Leonardo da Vinci, though mangoes and bananas may have been hard to get in Renaissance Florence. There are even several ultra-marathon runners who eat only fruit and claim it gives them special powers. But for many, this lifestyle is a modern form of eating disorder.

Juicing and detox miracles

'I look like I swallowed a sheep.' It was 2007, and a Sydney stock market trader named Joe Cross looked in the mirror and realised he was very fat. Without further delay he started a juice fast of sixty days. His aim was to lose weight permanently and get rid of an auto-immune disease which made him dependent on medication. 'I wanted to retake control of my life.' As a kid he had a big appetite and loved junk foods and sugary drinks, but was sporty and stayed slim. These habits continued into adulthood, and once for a bet he ate eleven Big Macs at a sitting. He regularly had four or five Cokes a day, as well as several beers to chase down his Chinese meal during working lunches. He had an addictive personality and had struggled with alcohol, and like his father he had been a keen gambler.

Over the years he had focused on being wealthy and successful and his weight progressively increased. He had tried all kinds of short diets and fixes before. He was even a fruitarian for a month, before relapsing into his old habits. His 2010 documentary film entitled *Fat, Sick, and Nearly Dead* was an apt description of his situation. At the age of forty, Joe weighed over 140 kg (308 lb) and was dealing with the rare auto-immune illness, urticarial vasculitis. He was at high risk of heart attacks and diabetes. According to friends, behind the jokey, rich, beer-drinking Australian this was someone on the brink of suicide.

The urticaria had developed ten years earlier out of the blue after a game of golf in California. It is a strange disease in which the small blood vessels overreact to all sorts of stimuli and produce histamine. Even in a large teaching hospital I have seen only a few cases, and it can also cause arthritis. It behaves partly like an allergy and partly like an auto-immune disease. When triggered by changes in heat, touch or even pollution, the skin comes out in big red blotches like nettle stings or horsefly bites.

Sometimes even a handshake could set him off, and the skin would immediately become red and blotchy. The body then reacts by sending fluid through the leaky blood vessels and the skin puffs up as in a major allergic reaction. There is no cure for the disease, although steroids and other immune-suppressants can lessen the symptoms. He was taking cortisone (steroids), which like for most auto-immune conditions helped a lot at first, but longer term it increased his appetite and exacerbated his weight problem (and it has many other side effects).

For sixty days Joe followed his chosen plants-only diet: he juiced for breakfast, lunch and dinner. He often had a fruit mix in the mornings, but his staple, which he called his 'mean green' mix, consisted of six kale leaves, one cucumber, four stalks of celery, two green apples, half a lemon and a slice of ginger. He carried his own juicer and generator with him wherever he went. He was not permitted alcohol, tea or coffee and had no other food or drink. The first three days, he said, were very tough, then he settled into it.

He deliberately did all this while travelling across America to see how people would react. He wanted to test his will against the ultimate junk-food country, he said, and be faced with continual temptation.

By the end of Joe Cross's juicing journey he had lost 37 kg (80 pounds) – that's around half a kilo a day – and his cholesterol levels had dropped 50 per cent. He said he felt great, pumped full of energy. This healthy glow continued when he started eating solid fruit and vegetables. He was able to gradually tail off, and under supervision his cortisone was stopped and his urticaria didn't reappear. He also managed to convert to his regime an even more obese truck driver – bizarrely, with the identical rare urticarial condition – whom he met in Arizona and who successfully followed his methods.

Joe Cross and his story have proved inspirational. Many people have tried his method, usually on two- to ten-day juicing fasts, and with differing levels of success. But he had certain little advantages: he was rich and single, could take sixty days off work, and he had a doctor and nutritionist to guide him with his health and his food. He also had a dedicated film crew in tow, although that wasn't always a help, he said, because he had to sit in the car while they were tucking into a McDonald's.

Importantly, he had the motivation of his illness and also the film about his personal trial. However, the real test was not whether he could complete the diet, but whether he could maintain his new healthy weight. Five years later things were still looking good. On a super-healthy vegetable diet with intermittent juicing he has managed to keep his weight off and has even lost some more.

Juicing and detoxing have taken off as methods to lose weight and 'reboot'. There are few or no scientific studies or trials of juicing, and most of the information comes from individual stories and promotional sites on the Web. It is certainly popular, and in many countries sales of expensive juicers and ready-made mixes have shot up. Clearly, large quantities of fresh fruit and vegetables are an ideal source of nutrients, most of which we can't get from meat and refined carbs. The concept espoused on many nutritional websites is that consuming juice rather than whole foods offers a number of advantages that put your digestive system in a relaxed 'holiday' mode. Is this true?

The nutrients are absorbed, we are told, without any effort and this process also gets rid of toxins in some way. These 'toxins' are never well defined, but some sites call them acids, dead cells or products of decay. The toxic products apparently accumulate in your cells, to dangerous levels. The 'mass of dead cells, toxins and acid then spills into your blood', causing inflammation. This leads to chronic sickness, a weakened immune system and an increase in most known diseases. Help is at hand, though. Juicing with the super-concentrated doses of super-nutrients combined with fasting eliminates your toxins, balances your blood pH and thereby purifies your body.

By now you will have realised that this kind of story which is repeated in many books and on websites is, sadly, complete mumbo-jumbo. The nutrients in fresh juices are of course good for you, but the pseudo-scientific claims widely made for detoxing the body by juice fasting are meaningless to any serious scientist or doctor. The ideas might have resonated in the Middle Ages when purges, leeches and blood-letting were all the rage, but our bodies (outside of sci-fi films) do not build up with excess toxins or spill over with acid or need a regular de-scaling.

Fasting to clean up your microbiome

Many people lose some weight through juicing but this could be another form of fasting. Although fasting has no common definition, it generally means eating or drinking between zero to 30 per cent of normal daily intakes. In 2013 a new fasting-based diet book was released which broke all sales records in the UK. *The Fast Diet* was championed by British TV presenter Michael Mosley.[5] Michael was making a BBC documentary on diets and fasting and tried out different regimes on himself. One of these involved several days of continuous deprivation on just a few hundred calories a day. This pure form of permanent fasting called Caloric Restriction (CR), is only for the ultra-determined. Michael managed it and lost weight, but felt this was too tough physically and psychologically for most people with normal lives.

After discussions with other scientists he tried out a more practical method which cut daily calories down to less than one-third of normal intake (500 calories for women, and 600 calories for men) for only two days a week, with five days normal eating. On the fasting days this could mean having an egg and fruit for breakfast, a handful of nuts and carrots for lunch and fish and vegetables for dinner, with plenty of herbal teas. Michael, although never what we would call fat, found he lost a stone (7 kg) in weight fairly easily over the five weeks and could eat and drink what he liked on non-fasting days. His body fat percentage surprisingly dropped even more rapidly from 27 to 20 per cent.

I tried the diet myself for a couple of weeks, just to assess its practicality. On my busiest days at the hospital, when I had no time to think about meals I found it easy knowing that I could, in theory, eat what I liked the following day. Like others, the next day after fasting I strangely had no extra urge to devour a full English breakfast and felt virtuous and somehow healthier. Michael and his team concluded that the most plausible reason for the metabolic and weight improvements were changes to a hormone IGF-1. This hormone, when switched off, has been shown to have anti-ageing properties in some animals and to reduce cellular stress. In humans, fasting produces an initial drop in blood IGF-1 which quickly returns to normal. But most of the good evidence of IGF-1's benefits comes from worms and flies; studies in rodents are less clear. The benefits may come at the cost of reduced development and a lack of energy and sexual interest.[6] This may only be worthwhile if you live for ever.

The results of intermittent fasting appear to be more significant than simply the effect of a reduction in calories and a temporary decrease in IGF-1 levels. A 2015 review looked at six small studies of intermittent fasting and found on average people lost 9 per cent of their starting weight after six months, which is comparable to stricter diets, but with a much lower drop-out rate (20 per cent).[7] There are no recent longer-term studies in humans, but in Spain in 1956 – under General Franco's regime and before ethics committees existed – a Spanish residential home for the elderly was used for an unrepeatable and little-known study. The home was run by a tough group of nurses from St Joseph's in Madrid. The 120 long-suffering residents (who were probably not chubby to begin with) were divided into two groups. One group had all their meals unequally distributed so the calories varied on alternate days: one day they got only 900 calories (a litre of milk and fruits), and 2,300 calories the next. The other group got the same average of 1,600 calories every day, which was considered normal for their size and age. After three years of this regime the alternating group had half the rate of death (6 vs. 13) and half the number of days spent in hospital for flu, infections and other problems.[8]

Humans have, of course, been intermittently fasting for thousands of years – in the name of religion. In fact every major religious group, including Christians, Muslims, Jews, Hindus and Buddhists and many more, have fasting as a part of their culture and discipline. It was clearly hard to start a new religion without some change in diet and a bit of healthy fasting. Most of the religious dietary practices adopted were linked to a potential health benefit, so fasting must have been given the same status.[9]

Many animals fast naturally in the wild. Hibernating animals such as squirrels have very different seasonal gut microbes when you examine their droppings before or after hibernation, and the change seems to be beneficial. Their levels of diversity are greatest after two weeks of re-feeding in the spring when their healthy butyrate levels go up.

Hibernating microbes living off scraps of the gut lining proliferate, and ones that depend on food disappear.[10] Burmese pythons grow up to 18 feet long and have a more varied diet than mice and pigs but never know when the next meal will come along. This means they have long fasting periods, which can last over a month, when their stomach shrinks down. Intrepid researchers examined python droppings and found similar microbial changes to the fasting squirrel in the microbe community and then large shifts one or two days after a tasty meal.[11] It appears that all animals have this natural relationship with fasting and their microbes.

Experiments on lab mice have shown that if you disrupt their normal nocturnal pattern of eating and make them eat more like humans their gut microbes become less healthy and they lose the natural daily cyclical pattern of change in their microbes. If you then compress all their eating into short 6–8 hour periods with 18-hour gaps between meals, they become even healthier, leaner and resistant even to high-fat diets.[12] The major microbe that loves fasting is called Akkermansia and it snacks off our gut lining and cleans it up, strangely improving the diversity of the other species. However, if we go for too long without food the microbe can damage the gut lining and cause problems. In our pilot studies for the American and British gut projects the greatest beneficial microbial changes were seen in the fasting groups with major increases in diversity. One small intervention study of obese patients showed major improvement in microbiome diversity and Akkermansia after a week of mild fasting.[13]

So if fasting short term is good for our microbes, should we be altering our meal times?

A healthy start to the day – or is it?

A popular nutritional assumption has long been that we should eat at regular intervals to avoid metabolic disturbances and compensatory overeating later. Another is that you should space meals out evenly to aid digestion. The idea of deliberately fasting may seem odd to many people, yet most of us fast regularly for 10 to 12 hours overnight without any problems. So why can't we also fast in the day for more than 4 to 6 hours – without complaining of a 'hypo' and reaching for a chocolate biscuit? The dogma in the Anglo-Saxon world about always having to have breakfast to maintain metabolic health and reduce overeating is even stronger. Over a third of the populations of Spain, France and Italy skip breakfast completely although a few people would have a quick espresso at the bus stop. They seem to be quite healthy and may not have much to eat most days until 2 p.m.

The fact is that we have been misled by indoctrination from breakfast cereal companies, compounded by a whole series of poor-quality cross-sectional studies. These reported an association between people who regularly skipped breakfast with obesity, poor glucose control and overeating later in the day.[14] The problem with these studies (as we have seen before) is that they are small-scale, cross-sectional associations focusing on the habits of obese patients who were already metabolically abnormal. Five recent studies have accounted for body weight while performing actual trials of a few weeks of skipping breakfast. They all showed no increase in weight or total calories in the breakfast skippers. In fact, there was a slight reduction in total intakes with no change in metabolic rate whether you were fat or lean, and most of the studies actually reported slight weight loss in breakfast skippers.[15] Evidence that most children suffer from lack of breakfast is equally weak and no proper trials have been done.[16] So compulsory breakfast is another diet myth that should be buried.[17]

Some people still feel they really need their breakfast: this is partly cultural and partly genetic. Our twin studies have shown a clear gene influence on whether you are a morning person or an evening person, and these differences in circadian rhythms undoubtedly affect the times when you prefer to eat.[18] We should let our bodies not dogma or guidelines determine our choice of breakfast and mealtimes.

The three-meals-a-day routine is actually a surprisingly modern invention – only coming to the west in the Victorian era. We don't know for sure but suspect our Palaeolithic ancestors only had one main meal a day. The Greeks, Persians, Romans, and early Jews all ate one big meal per day, usually in the evening to celebrate the day's work. It wasn't until the sixteenth century in England that two regular meals became more common, but only for the well off. A contemporary proverb proclaimed this was the healthy new recipe for long life: 'To rise at six, dine at ten, sup at six and go to bed at ten, makes a man live ten times ten.' The habit of only two meals a day among even the rich upper classes is made clear by the Countess of Landsfeld, who in 1858 described her companions eating habits: 'after this meal comes the long fast from nine in the morning till five or six in the afternoon, when dinner is served.'[19]

The evidence clearly shows that increasing our fasting periods, whether by the 5:2 method of having two hungry days eating 25 per cent of our normal calorie intake, or milder variants such as cutting out or compressing some meals, could actually be better for us – even if we consumed the same number of calories daily. One shorter version of the Spanish nursing home study gave volunteers the same food content and calories either as one giant single meal or as three small divided meals. The study lasted for eight weeks and then after a break the volunteers swapped over to the other regime. There were no significant differences on heart rate, body temperature, or most blood tests. However, the single meal produced more feelings of hunger, slight increases in blood lipids, but significant reductions in body fat and in the stress hormone cortisol.[20] So although you may feel hungrier by skipping two meals, there is no evidence it will do you any harm, and it may have even have metabolic benefits for you and your microbes. Their powerful circadian rhythms are important for our immune system and health, and are helped by fasting.

Superfoods and super microbes

'Superfoods' form the basis of a massive growth industry. People are willing to pay big money for what they perceive to be the most nutritious ingredients, or for an exotic nut or vegetable that their friends have never heard of, especially if it is reported to have magic healing powers. Of the few superfood studies that make it to bona fide scientific publication, most of the 'amazing results' are shown only in test tubes or, if rarely, in rats fed huge amounts of the pure compound. Only a tiny fraction have ever been studied properly in humans, at the correct doses or with normal food – and then only very short term. Commonly cited examples of superfoods are pomegranates, blueberries and the highly touted acai and goji berries with their potential antioxidant properties, which, despite the hype, have not been shown to be different to other berries. Even boring beetroot has been promoted for its apparent nitric-oxide-altering properties.

Other trendy superfoods are more exotic. They include freshwater Asian marine algae like Chlorella, which is very green and apparently has immune-disease-, diabetes- and cancer-fighting properties and costs 'only £90' for enough to last a month. Blue-green algae, like Spirulina (which I was given in the form of a slimy soup recently), is another 'immune-busting' superfood with plenty of protein and vitamins. Unfortunately it costs, per gram, thirty times more than meat. Spirulina is formed by groups of microbes that stick together on lakes and are actually a type of ancient bacteria – cyanobacteria. These microbes were responsible for providing the earth's atmosphere and evolved into chloroplasts in leaves, so we could consider them ancient probiotics. Some of these microbial species that we thought only existed in water we actually found living in the guts of our twins.[21] This reminds us how much there is still to learn of the microbes we share our bodies with.

Like other bacteria that can produce vitamins, Spirulina produces vitamin K and a form of B12 that has been heavily touted by vegan websites as an alternative to meat. However, there is no evidence that it has the same key properties or benefits as real B12.[22] As with seaweed, unless you have the right gut microbes and their 30,000 special enzymes, you will not be able to extract the nutrients from the more exotic marine superfoods.

Superfoods may seem a fun concept, but they are also a marketing con, as virtually every fresh fruit and vegetable is a superfood. Trendy examples include chia seeds, hemp, quinoa and wheatgrass thanks to their high fibre and vitamin content. They are all packed with hundreds of different chemical compounds that people can dream up long lists of attributes about. Some consider yoghurt, eggs and most nuts to be superfoods, and from the earlier chapters of this book we could add traditional cheeses, olive oil and garlic – the potential list is endless. One book on the market lists the latest top 101 superfoods.

What is becoming clear is that many foods don't work nutritionally as well in isolation as they do when combined together. Spinach and carrots are good examples, as both contain the nutrient carotene that is better absorbed in the presence of the fat in olive oil dressing.[23]

This is yet another reason, if you still needed any, to avoid reductionism in our food choices. Sporadically feeding yourself large amounts of a few selected 'superfoods' is likely to be much less effective than regularly eating a wide variety and diversity of plants and vegetables.

Juicing long term with too much fruit can be dangerous, as it produces large amounts of fructose and calories without enough accompanying fibre, which among other benefits dampens down the speed of the sugar absorption. Dentists have noticed recently increases in new cavities in healthy, over-keen twenty-something juicers. Instead, adding a regular multi-vegetable juice to your diet appears much healthier. It can be prepared cold or briefly heated as soup. Studies have shown that people are more willing to eat a wide variety of vegetables as soup than if served naked on a plate. If the soup is thick and the vegetables only lightly cooked, it also slows down digestion and can send signals of fullness to the brain and lower intestines. The other advantage is that nutrients are not lost in the water, and most people in normal health can eat many more plants as a juice or soup than raw or by traditional cooking. Using juicing machines that preserve the pulp and fibre and so many of the nutrients seems sensible if it doesn't become an obsession.

Be kind to spinach haters

The average Western diet is deficient in fibre, and in carbohydrates from vegetables and fruit. As a consequence it is likely that the average healthy person is actually abnormal in terms of their gut microbes, as the gut community are being starved of accessible carbohydrates.[24] It doesn't help that many people have strange aversions to many vegetables – we discussed a few cases earlier. One study of babies aged eight to eighteen months found real aversions to touching or eating green plants, similar to the natural aversion to snakes or insects.[25]

This green aversion manifested, for instance, as a hatred for school spinach, may have an evolutionary basis as a means of stopping children randomly picking up some toxic green plant and eating it. Genes, as always, play a role, and as all parents know, some kids have stronger aversions or food avoidance instincts than others, particularly for new foods. This so-called neophobia often continues into adulthood as fussy eating. Such individuals are less likely than others to try any new food or taste, and end up on more restricted low-nutritional diets. We found in our adult twins a very strong heritable and persistent fussy-eating component.[26] Changing the particular food's physical form by using tricks like juicing may be one way of reducing these aversions; and introducing bright non-green colours could help.

In the modern world we should treat as a superfood any plant or vegetable that assists our microbes in their work. But rather than focusing on single ingredients and spending our twilight years eating Spirulina soup three times a day, we should be thinking more of superfood communities. The wider the range, the better. Whether you achieve this via soups, juices or raw or cooked foods is probably not so crucial as long as variety is maintained. As well as what we eat, the timing and space between meals may be more important for us and our microbes. Gut bacteria along with all animals need to have regular patterns of work and long rest periods to function optimally. As well as the nutrients and polyphenols, many real foods contain fibre that makes them tough to digest and give our microbes a good workout, and this we were told is good for us.

12

Fibre

One pioneering Irish doctor called Dennis Burkitt had a major influence on our modern relationship with fibre. Burkitt was a legend, and probably the last of his breed of zealous explorer-scientists that the British Empire produced.

After his training in medicine and surgery he was sent during the Second World War to East Africa. Later, in his forties, he answered 'God's call' to go and work as a missionary and doctor in central Africa. There he spent several years, and covered ten thousand miles travelling around the country visiting small hospitals and health centres performing operations and preaching.

To pass the time as he travelled, he drew maps on which he plotted the prevalent diseases. For instance, he discovered a lymphoma which affected children that only occurred in the malaria belt, and righty predicted that it was due to an infection, a virus, which could be treated. He also plotted the dietary and bowel habits of the natives, and in the 1970s came up with another theory.

I remember vividly a lecture he gave while I was a student at the London School of Hygiene, where he showed his travel snaps. These were mainly pictures of impressively large African turds taken in exotic locations. The Kalahari bushmen regularly produced specimens that weighed a massive two pounds on average, compared to the average 'civilised' European offering of 4 ounces. And on his maps he had linked the quantity of poo produced with the amount of fibre in the diet and the absence of Western diseases.

His notion was that the main benefit of fibre was as a bulking agent and stool softener that would speed up transit though your bowels so toxins couldn't get back into your body and cause cancer. As a bonus, fibre also mopped up the fats that cause heart disease, and prevented haemorrhoids and varicose veins. His observations were astute and ahead of his time. He also criticised the modern trends of refining carbohydrates, denuding them of their high-fibre covering, and the eating of white flour.

He was convinced that the modern toilet position for defecating was bad for us and that a lack of fibre caused colon cancer. Subsequent studies failed to support this colon-cancer theory, and the latest studies have shown that, surprisingly, even constipation, which certainly causes other problems, is not a cancer risk factor, although high-fibre laxatives may be protective.[1] Nevertheless, thanks to Burkitt's missionary zeal, fibre had become trendy and part of our vocabulary.

'Dietary fibre' is the general term for the bits of food that can't be digested. It used to be thought of as completely inert, with no real effect or interaction with the body except for its mechanical properties. But fibre comes in different forms and can be soluble like oats, beans and fruits, which are fermented in the colon, or insoluble like whole wheat, nuts, seeds, bran, fruit skins and many legumes and other vegetables such as green beans. However, even the insoluble type is not totally inert and can be fermented by bacteria to produce gases and other by-products. In general, fibre absorbs water and speeds up transit through our intestines. While most people accept that it is good for us, there is a lack of consensus about why. Prebiotics could be one reason.

Promoting the benefits of fibre in the diet was big business well before Burkitt's work, and has an origin dating back to the time of the ancient Greeks.

All fibre is a form of carbohydrate, and so in theory provides four calories per gram in energy. But most is not absorbed, and food labels can be deceptive and confusing. In the 1980s at the height of the low-fat movement, the US population in particular were urged to eat lots of oat bran. Fibre has since 1993 been allowed as a health claim on US food labels, depending on how much fat is also in the food. These health claims haven't helped change habits much: most Americans (and Britons) eat less than most other countries, consuming only half of the current modest recommendations (18 to 25 grams per day), and kids eat even less.

At the time the Food and Drug Administration approved a health claim for fibre, the observational data was in fact pretty shaky. A recent meta-analysis of these twenty-two reports on heart disease found quite big differences between studies (which is always a worrying indication of their quality). Nevertheless, it concluded that fibre was overall beneficial. The estimates were that you could reduce the risk of heart disease by 10 per cent for every extra 7 grams of fibre consumed.[2] Nearly the same protective effects were found in reducing overall mortality, in seven studies of nearly a million people.[3]

In most studies whole-grain fibres were considered to be a major health factor, although benefits were primarily seen for other plant and vegetable sources. The data is weakest for fruit fibre. To increase your daily dietary fibre by 7 grams is not that onerous, requiring one extra portion (100 grams) of whole grains, one portion of legumes (beans or lentils), or one to two portions of green vegetables or four pieces of whole fruit (but only if you eat the skin).

As just noted, oat bran for breakfast was a big marketing sensation in the US in the 1980s and 90s, after early studies showed it reduced cholesterol levels dramatically. It was also promoted as miraculously reducing blood pressure and diabetes risk. The marketing gurus lost no time, and the *New York Times* reported in 1988 a frenzy over oat-bran muffins, as stores ran out and it became 'the croissant of the 80s'. Conflicting studies followed which cast some doubt on the amazing health benefits claimed, until a more recent meta-analysis of sixty-six oat studies clarified the evidence. This showed no effect on diabetes or blood pressure but confirmed a consistent beneficial effect on blood cholesterol.[4]

Unfortunately, the benefit of oat bran turned out to be tiny, with only a modest 2 to 4 per cent reduction in cholesterol unless you had very high levels to start with. To achieve this meagre reduction you had to eat an enormous Daddy Bear-size bowlful (three packets) a day. The muffins were worse. Packed with extra calories and a high fat content, any modest health benefit of the bran was overwhelmed. Nevertheless, the scientists of the 1990s were confused about why bran would have such an effect at all. Thirty years later, microbes provide the answer.

Prebiotics and microbe fertilisers

Prebiotics are components of food that are closely involved with our healthy microbes. While not all fibre is prebiotic, all prebiotics are by definition non-digestible fibres. It is not surprising that when you measure levels of prebiotics (like the common prebiotic substance inulin), they mirror the fibre content of our diets. Data from my King's College colleague Kevin Whelan shows a good correlation between inulin and fibre in the UK, indicating the two are closely linked. This connects the new science of prebiotics with the old-fashioned concept of fibre.

One important way in which our food and microbes interact is via prebiotics. Whereas *pro*biotics are selected microbes that benefit the health of the host, *pre*biotics are the constituent parts of foods that act as fertilisers for the microbes in the colon. These largely non-digested fibres allow beneficial microbes to thrive, and they come in several forms. The first prebiotic we encounter in life comes handily packaged in breast milk and is called an oligosaccharide, which is a complex group of tightly bound sugars.[5]

Many prebiotics are referred to as resistant starches, as opposed to highly refined (that is, broken down) starches like rice or pasta that are easy to digest and release glucose. It has been crudely estimated that a healthy person needs about 6 grams of prebiotics a day to keep both their microbes and themselves healthy.

Several types of prebiotics are well established scientifically, and there are many others that have potential effects but lack real proof; and all forms are of course marketed equally on the internet. The well-known ones include the major player just mentioned, inulin (not to be confused with insulin), as well as oligofructose and galacto-oligosaccharide. You will come across them more and more, as they are being increasingly used by the food industry as additives; when combined with probiotics they are called synbiotics. A recent innovative addition are so-called optibiotics, where a probiotic with special health properties is combined with a prebiotic that has been specially selected to fertilise that microbe preferentially and so maximise its chances of survival. Tests on a cholesterol-lowering combination are ongoing, and we are likely to see many more examples of these.

These prebiotic compounds come from natural ingredients such as chicory root, Jerusalem artichoke, dandelion greens, leeks, onions, garlic, asparagus, wheat bran, wheat flour, broccoli and bananas and some nuts.[6] The percentages of the prebiotic inulin in each one vary enormously, from about 65 per cent in chicory root to only 1 per cent in a banana. Most increase their active amounts if dried, but lose half if cooked, so you'll need to eat more if you don't like your food *al dente*.

To achieve your prebiotic quota (6 grams) would mean eating each day half a kilo of bananas (ten), or alternatively only a teaspoon of diced chicory root or Jerusalem artichoke. Grains and bread too, surprisingly, contain about 1 per cent of inulins – rye bread slightly more than others – and even 'plastic' sliced white bread contains reasonable amounts.[7] The decline in the consumption of fibrous vegetables has made bread the estimated main source of prebiotics and fibre in both the US[8] (2.6 grams daily) and the UK (4 grams). Non Anglo-Saxon Europeans, particularly those on Mediterranean diets, have fibre intakes around three times higher.[9]

Garlic breath cures the common cold

Garlic, as well as being an excellent source of polyphenols and vitamins, is a first-class prebiotic that used to be a major discriminator between the cuisines and habits of northern and southern Europe, and has been used for millennia in Asia. Before the 1980s it was uncommon in the UK, and I remember in the early 1970s as a kid the novel sight of a Frenchman with an ostentatious moustache and dressed in a striped blue jersey and black beret cycling around the London suburbs festooned with strings of exotic and suitably expensive garlic and shallots. I also remember on school trips the overpowering smell of garlic breath on the early-morning Paris Metro – though I probably wouldn't notice it as much now because the cosmopolitan experience of the London Underground is not so different.

In 1976 Marks & Spencer nervously produced the UK's first ready-meal and, controversially, it had garlic in it. Going by the exotic name of Chicken Kiev, it is now, not surprisingly, a dish whose origins are still being hotly disputed between Moscow and Kiev. It was a massive hit, and Brits have been steadily getting used to the taste ever since. McCain the potato company have even launched roasted garlic wedges as a snack, which would have been unthinkable in the UK just a few years ago.

Sometimes I come across southern Europeans who can't stand garlic, which I find very strange, as I've always assumed all southern babies are fed it from birth and so get rapidly used to the strong taste. We conducted a study in over 3,000 of our twins to confirm if garlic eating in the UK population was mainly due to cultural exposure. Contrary to our expectations, the results showed a strong genetic component (49 per cent) in whether people ate garlic or not, with only a negligible influence coming from the family environment.[10] This suggests that genes for taste receptors, particularly bitter flavours, are important, at least in the UK. These genes may be rarer in the southern Mediterranean where fewer people dislike garlic.

Garlic on its own has been hyped as having many health-giving properties, such as preventing and curing colds, cancer and arthritis. Although I have published some of these studies myself on arthritis, the results, though intriguing, have yet to be convincingly proven and could just be markers of a healthy diet or lifestyle.[11] Garlic has a strong tradition in Mediterranean countries as a cold remedy. I once tried a Tuscan remedy for cold prevention. At the first symptoms you take three cloves of raw garlic and a full bottle of Chianti. The results were amazing. The next day I woke with garlic breath, a bad hangover and the predictable cold symptoms. I later was told that I should have taken it *before* getting the cold.

A recent independent Cochrane Review looked at the eight trials of garlic and colds. Sadly, they found only one worth assessing. The UK investigator randomised 146 people to *Allium sativum* (garlic) and the other group to a placebo. After twelve weeks, the garlic eaters had experienced three times fewer days of symptoms than the dummy group.[12] The downside is that the dose was equivalent to over eight cloves of garlic a day, which could be a problem for some people. However, I can share this tip – that garlic breath can apparently be relieved more quickly by ingesting a mixture of parsley and yoghurt, which together provide an excellent microbial air freshener.

Garlic's effect on lowering cholesterol and improving lipid profiles does seem more likely to be real, based on multiple replications of the results and a meta-analysis of several randomised studies.[13] The benefits of garlic may, though, depend on what else you are eating, as well as which microbes are already in your colon.

Spring-cleaning your guts

I recently discovered the practical problems of eating natural prebiotics after 'volunteering' to have a colonoscopy at my own hospital. This has not yet become a widespread British hobby or bucket-list item, and I was doing it for several reasons. I had never had one, and my US colleagues, who seem to have them as often as they have haircuts, thought it very backward of me. Many countries now advise colonoscopy as a routine screening test for colon cancer for anyone over the age of fifty. Colon cancer is one of the most preventable cancers in men. I also try to practise what I preach. I am planning a colonoscopy study of twins, and I generally try out any invasive tests before asking our twin volunteers to do the same. At the same time I wanted to test my microbes, to see how they would react to being cleaned out, for them the equivalent of a giant tsunami invasion.

The days just before the test, when I had to cut down on fibre, were fine. Some people don't eat enough to notice, but I held off the fruit and vegetables and whole grains. The fasting was fairly easy as I could still drink fluids. I was warned to carefully plan the next few hours after taking the sachet of strong laxatives. I shouldn't be at work, definitely not on a crowded train, and if possible I should stay within ten yards of the toilet. I thought that was a bit excessive.

It took a while to work but, to spare you the details, I can say I was glad that for once I had heeded the advice. Twenty trips and a loo roll later I was finally cleansed and empty and ready. The actual procedure was for me painless and interesting, probably because I had one of the best endoscopists in London, Jeremy Sanderson, at the business end. He gave me a tiny dose of 'happy juice' as he called it, which is basically a short-acting Valium-type drug, and off we went.

I watched it all in colour on the large TV screen at the side of the bed as he worked his way around my slimy, gleaming intestines, taking eighteen tiny biopsies for our studies. I wanted to go back to work but was told to take the day off and relax, as the happy juice might have made me do odd things. In this laid-back state I started thinking about my billions of poor hard-working microbes that I had earlier flushed down the toilet.

It may have been the drug, but I became quite emotional. I felt sorry for those faithful microbes that I had being trying to carefully cultivate for the last year. I knew from studies of antibiotic-treated patients and others who'd had colonoscopies that over 99 per cent are wiped out. Gritty survivors hang on in unusual places, and if, like me, you still have one, they can hide in the haven of the appendix – could this be its long-lost purpose, perhaps? They can also collect in the caecum, a corner of the colon that always contains some fluid, a smelly oasis in the desert. Microbes can lie low, too, in the tiny crevices of the intestinal wall, clinging on by forming biofilms with each other – although no one quite knows how they resist so well the tidal waves rushing through the bowel.

The three-day prebiotic diet

There are only a few reports following colonoscopy in humans, and the largest comprised fifteen patients. After a month, most recovered their previous microbial flora. However, three of them developed major changes for unknown reasons.[14] Anecdotally, when I ask my Gastro colleagues they say that every now and again some patients with mild colitis or IBS do report a miraculous cure after the clear-out. This is likely to be due to major changes in their gut microbes. Anyway, I wanted to give the survivors in my bowel a reward for their persistence and decided to go on an intensive three-day pre-biotic diet.

First, I needed supplies. No one had any Jerusalem artichokes as they were out of season, so I had to substitute globe artichokes. Despite the similar name, Jerusalem artichokes are sunflower roots that look like potatoes and are called 'sunchokes' in the US. They are affectionately known as 'fartichokes' by the Brits, some of whom appear to suffer special – possibly genetic – side effects. I found chicory (a.k.a. endive, the unopened bitter leaves popular in northern France and

Belgium), but not the high-inulin-content chicory root itself, which would have been ideal. Dandelion leaves were never going to be easy as I don't live on a farm, and dandelion wine didn't count.

The rest was not a problem – garlic, onions, leeks, asparagus, broccoli, etc. I also added flax seeds (linseed), pistachios and a variety of other nuts. I made a giant salad, finely chopping the ingredients and adding a few spinach leaves, bits of tomato and parsley for seasoning, plus of course extra virgin olive oil and balsamic dressing and some whole-grain rye bread. It tasted good to me, though having this raw food three times a day produced a few side effects like a bit of wind at both ends, and no one wanted to kiss me. Just after the purging and colonoscopy I had felt great, which is a common report after many purges or fasts, and something that we don't fully understand.

But was this effort worth it for my microbes?

The British Gut Project lab had tested the results and found that the microbes just after the wash-out, although diminished, were pretty much the same mix of species as before the laxative maelstrom. However, one week after my rebirthing diet, thanks possibly to the prebiotics and the impressive survival and reproductive skills of my microbes, I had more bifidos and a richer variety overall than I had before, including some new species that had grown enough to be counted. The prebiotics had apparently done some theoretical good, in me at least, but a study of one is far from proof.

Do prebiotics really work, in properly conducted trials, versus a placebo? Most scientific prebiotic studies use doses of the actual chemicals rather than chopping up and eating the foods that contain them, as they are easier to measure and quantify. Between 5 and 20 grams of inulin per day is most commonly used in trials, and some trials use oligosaccharides while others use a combination. To qualify as a proper prebiotic, there has to be as a minimum standard a significant increase in bifidobacteria. There are many prebiotics that haven't been formally tested in humans, though this doesn't stop them being marketed and sold.

A recent meta-analysis summarised the results of twenty-six trials exploring prebiotics' effect on weight, which involved a total of 831 people. The overall quality of the trials was poor. As well as being small-scale, they were also of short duration, lasting a few days to a maximum of three months.[15] While most studies showed an increase in beneficial microbes, the effects on weight loss were not obvious or consistent. Only five studies actually tested obese volunteers.[16] Despite these problems, there was a remarkably consistent 40 per cent increase in self-reported feelings of fullness after meals, as well as a reduction in blood insulin and glucose levels.

The likely explanation behind these consistent findings is that prebiotics provide the fertiliser to increase the numbers of our healthy microbes (like bifidos) and to effect subtle changes in other mysterious microbes that we haven't yet quantified. These microbes then produce short-chain fatty acids that have a range of important effects on the body.[17] The most important of these is butyrate, which plays a part in releasing hormones in the gut to help suppress hunger and reduce the rise of glucose and insulin that otherwise put fat into storage mode. It is unlikely to be a coincidence that butyrate is also most successful in toning down the immune response and creating a state of calm tolerance.

If eating lots of prebiotics isn't your thing, you can skip the vegetables; and you can easily buy synthetic butyrate supplements in the US and online, with full FDA blessing. But this comes with several caveats. First, it won't have been properly tested in humans;[18] second, it wrongly assumes that butyrate on its own has the same effect as when surrounded by all the other natural chemicals; and finally, you should know that the natural butyrate substance is what you smell when butter goes rancid and what gives human vomit its distinctive aroma. Greenpeace have even used it as a stink bomb to throw at whalers.

Deadly grains and Franken-wheat

We have talked about the key role of whole grains in the context of Mediterranean-style diets, and as a source of prebiotics as well as of fibre. There is a growing movement that, by contrast, believes that whole grains are inherently unhealthy. Others go further, asserting that all grains are poisoning us and are to blame for obesity and all Western illness. Responsibility for this phenomenon lies largely with some bestselling books that have appeared in the US such as that by Dr William Davis, a US cardiologist whose website proclaims: '*Wheat Belly* was the original book that turned the nutritional world topsy-turvy and exposed healthy whole grains as the genetically altered Frankenwheat imposed on the public by agricultural geneticists and agribusiness.'[19] According to this doctrine, we are all to some extent intolerant of, or allergic to, this innovation of ten thousand years ago, and we have never been able to adapt. We should give up all foodstuffs containing grains, or risk the dire consequences.

Now it is certainly true that some people are intolerant of a protein component of wheat and of most other grains known as gluten, which sticks the molecules together (*gluten* means 'glue' in Latin) and gives bread its elastic properties. It is this protein that causes the auto-immune condition coeliac disease, which triggers a shrinking of the finger-like projections (villi) of the intestines' lining as well as serious digestive problems and malabsorption. Contrary to popular perception the disease is pretty rare, proven cases occurring only in 1 in 300 people, while the potential for getting it (along with blood antibodies) is only about 1 in 100.

In the UK and the USA, ten times more people think they have the disease than actually do, based on blood or gut changes; and, ironically, only about one in ten with the real disease gets properly diagnosed. The upsurge in interest in gluten has led to a big increase in fast-food chains and restaurants offering gluten-free products. This is a market worth $9 billion in sales in the US alone, and showing 20 per cent annual increases. It is big business, and the trend is spreading to Europe. Even small corner shops are stocking gluten-free cakes and breads now, while saying goodbye to soy and quorn.

In his book Davis claims as a justification of his diet plan that most coeliac sufferers lose weight when they go on a gluten-free (GF) diet. In fact, the opposite is true, and as a doctor I have only ever seen skinny coeliac sufferers who can't absorb food properly. In the study that Davis quoted, more than three times as many patients gain weight than lose it on a GF diet (95 as against 25), even those already overweight.[20]

Anyway, despite the lack of scientific evidence, the diet, the book and the recipes hit a chord with the American public, already suspicious of the food companies, and they have been a great success.

Book sales have totalled over a billion dollars in the US and spawned many other restricted-gluten or semi-vegan diets and cookbooks. Some people undoubtedly lose weight, but in most this may have nothing to do with the gluten. As we have seen many times before with specific restrictive diet regimes, they are excluding many food groups and dramatically reducing snacking opportunities.

Cutting out wheat, barley and rye from your diet could be OK if you replaced them with other healthy vegetables, but this often doesn't happen. For many people their diet becomes restricted to odd items like gluten-free cheese pizza and gluten-free beer. Consequently, they may lose out on valuable sources of B vitamins, fibre and prebiotics and see a big deterioration in the variety of their microbes.

Real coeliac sufferers are highly motivated to stick to the diet, as they become profoundly ill after ingesting even a small amount of gluten. But for most normal overweight people, without symptoms, to cut out all grains for life is obviously much harder. Small amounts of gluten are hard to avoid because they are routinely added to most processed foods and sauces to bind them and improve the texture. Avoiding processed foods may be the only real benefit of GF diets for most people.

Saliva mutations and vegetable evolution

The concept of the benefits of grain-free diets derives from the assumption, already mentioned, that up until nine thousand years ago our ancestors had never eaten them. So we haven't had the time to evolve the genes and the mechanisms to digest grains properly, causing a toxic allergic reaction that inflames our intestines and causes obesity and other problems. What is more, the high calorie count of all those grains we ingest can't be a good thing. We have already discussed how human genes mutated around seven thousand years ago to enable many of us to drink raw milk, and just a few years ago a similar notion concerning the supposed toxicity of dairy foods was very popular.

So why are we thought by some not to be flexible enough to safely adapt to eating grains or other starches in the space of nine thousand years? There is now evidence that we *can* adapt. The enzyme amylase is the one to concentrate on here. As with many other mammals, amylase is in our saliva to break down carbohydrates and in the pancreas for releasing into the small intestine.

A group of American geneticists had the bright idea of looking into how many copies of the gene there were in different populations around the world following very different starch diets.

Starch is part of all plants, and is virtually the only carbohydrate in potatoes, pasta and rice, as well as being found in wheat and root vegetables. Starch can be relatively easy to break down, as in cooked potatoes, or difficult (resistant) as in raw vegetables. After mastering cooking all those millennia ago, we began to eat these tough root vegetables that previously, when raw, were either not worth the nutritional effort or potentially toxic. We then started to cultivate them in large quantities in different parts of the world.

The American researchers compared the genes of rainforest Africans and Arctic Siberians with those of grain-eating Europeans and Africans and rice-eating Japanese. Sure enough, they found large differences in the number of copies of the gene.[21] Tribes on the traditional diet carried many fewer copies than those now on a starch diet. The more copies of the gene you had, the more amylase enzyme you produced and the better able you were to break the starch down.

We share over 99 per cent of our genes with monkeys. However, monkeys who mainly eat fruit and occasionally meat have no copies of the amylase gene. As with milk drinking, we in the West have adapted rapidly to our high-starch environment, probably because this gave us a major evolutionary advantage. The current theory is that having extra copies of the gene may have protected children from dying from diarrhoeal diseases as they could still get energy from starch.

With colleagues from Imperial College London we took this story further, using our twins. We worked out how many copies of the amylase gene each twin had and compared the numbers of copies with their body weight. This gave us a clear result – but not in the direction I had expected.

Those with the most gene copies and therefore the most amylase enzyme – meaning they could in theory digest the starch more easily – were the leanest; and those poorly adapted and with fewer copies (and poor digestion) were the fattest.[22] I had expected that better digestion would signal that they were extracting more calories from carbs and so weight gain would be the result, not the reverse.

This was a puzzle I was determined to crack and I thought that microbes could give us the answer, as what happens to food early on in the digestion process dramatically alters the structure of the food that enters the colon, where the interactions with our microbes occur. As is often the case, studying twins could point us in the right direction.

Carbohydrate-eating gene differences

Linda and Frances are sixty-eight-year-old twins taking part in our research. They don't often get mistaken for twins as Linda is 12 stone and Frances 8½ stone and being non-identical, like ordinary sisters, they only share half their genes. Linda was only 6 ounces heavier at birth but had always been the bigger of the two as long as they could remember. At the age of sixteen, seeing her sister attracting boyfriends, she went on the first of several diets – initially a semi-starvation diet that briefly got her close to her sister's weight, but she soon slipped back. They lived together till their mid-twenties. They had similar tastes and always consumed the same foods and drinks and had the same portion sizes. Linda, despite doing more exercise and sport, just kept on gaining weight while Frances didn't.

'We always knew that we had different metabolisms,' explained Frances. 'I often felt guilty as it wasn't fair for her and I could always eat what I liked.' Linda commented: 'I wasn't bitter or angry but did keep trying to lose weight by dieting and exercise, as I could see that she was getting all the boyfriends.'

Over the years Linda has tried many diets. The most sustained was the Atkins Diet, for six months, which worked well until she got bored with the repetition and stopped. She also tried the cabbage soup diet, 'which also worked for a few months but had other side effects, as you can imagine. I've realised that for me diets are pretty pointless so I've stopped controlling my food too much, but I still go to the gym and play golf and tennis and have an allotment to stay healthy and have fresh vegetables.'

As part of the amylase research project we had extracted their DNA from their blood, along with a thousand other pairs of twins. We then measured the numbers of copies of the gene they both had – which, incidentally, is still a very tricky procedure, and as each test is rather imprecise we did it six times, then averaged the results. In their case the results were clear-cut. Linda had only four copies of the amylase gene and Frances had nine. For each copy of the amylase gene you are short of, your risk of obesity increases by 19 per cent. So for Linda, despite much the same diet, her risk was approximately double that of her sister.

'These results, now you have explained them to us, do make sense. I can see why my sister and I are different and do indeed have different metabolisms in how we react to food. It may be too late for us, but can we test our children?'

The results were so surprising that we spent a year checking them, and as with most diet stories ours made a big media splash. To put it into context, we had found a gene effect that was around ten times larger than anything found before, including the obesity gene, which was the first to be easily discovered (called FTO) but now looks much less important and is a useless predictor in individuals. The downside is that this gene effect is currently very difficult to measure accurately without incurring large costs.

A potato is not the same for all of us

The other major paradigm shift is that until now the vast majority of the genes discovered for obesity have been thought to act on the brain. This has led to the continued perception that it is just a question of greedy signals from the brain causing the weak-willed obese person to overeat. What we found is that metabolic effects (that is, those affecting energy) are ten times greater. Our signals of gene copies were very much harder to pick up, and there may be many more like the amylase gene for other food types that we'll discover in the future. Although the field is likely to change again, the latest work shows us that in contrast to our earlier beliefs, obesity genes (like FTO) may work by altering the size and type of our different fat cells (white, beige and brown) rather than purely on the brain.[23]

We also looked at the microbial and metabolic patterns of those of our twins who had the highest and lowest copies of the amylase gene and found major differences in some of the Firmicutes family (clostridiales), which have been associated with obesity. We don't have the full story yet, but so far it looks like the altered digestion of the starches in maladapted people triggers a change in microbial composition and the production of different fatty acids. This in turn could lead to more rapid increases in insulin on eating starch, which ultimately leads to predisposed people storing more fat and an increased risk of diabetes. This indicates that some people eating exactly the same bowl of potatoes or pasta will have a greater amount deposited as fat because of the effect of their genes on their microbes. So a potato is not a potato to everyone – to some people that potato, energy-wise, is like a double portion.

Finding more of these gene copies could help us in the future to divide people into different groups of food eaters. People like Linda who find themselves in a high-grain-eating environment but lack the right genes may be better off reducing the high-carb starches and eating fats instead, which their bodies and microbes may be better designed for.

The good news for the rest of us is that we can adapt to new foods and environments faster than we think. As well as the extra gene copies to produce more starch-digesting enzymes in areas with high-starch diets, we also mutated our genes to be able to consume cow's milk, and there are no doubt many other similar gene mutations we haven't discovered yet. We also know that exposure to different diets can actually change our genes (epigenetically). The human body is much more flexible than we ever believed. We are not mass-produced like robots – we are much more plastic, with the ability to adapt to our surroundings. This has been the secret of our survival and success on the planet and has enabled us to exploit all the diverse diets and environments available. This works well for the population as a whole, but individuals like Linda may not be able to adapt their genes or their metabolism and may need to alter their diet or their microbes.

Low FODMAP diets and bloated bellies

Most people with irritable bowel symptom (IBS) have heard of the low FODMAP diet. FODMAP stands for fermentable oligosaccharides, disaccharides, monosaccharides and polyols. The diet involves restricting intake of these carbohydrates, such as fructans and galacto-oligosaccharides, which are poorly absorbed by the body. This cuts out many foods including wheat, legumes, pulses, and some fruits and vegetables. It can result in a dramatic improvement in symptoms in some IBS sufferers, but the results are hard to predict. The downside to this diet is that the restricted foods contain lots of important nutrients, including fibre and vitamins as well as polyphenols, and there may be adverse effects on the health and diversity of many people's microbiome. It is therefore essential only to follow this restrictive diet with support from a registered dietician.[24] Once symptoms have been alleviated the trick is to slowly reintroduce a few FODMAP-rich foods to see if they are tolerated. The wide variation in the microbes of sufferers can make this unpredictable too.

The health benefits of the Mediterranean diet I mentioned earlier are probably not just to do with olive oil, red wine, nuts and dairy. The range and fibre content of other foods used weekly is large. We have mentioned tomatoes, onions and garlic as part of the basic sauce that goes on top of most of the grains, which usually take the form of bread, rice or pasta. But let's not forget the other legumes (including beans), chick peas and so on or the cruciferous vegetables that are commonly eaten. As well as containing polyphenols that have important effects on the microbiota, a lot more fibre needs to be consumed. We currently eat much less than we should and we cannot afford to permanently cut out these major and varied sources of fibre and nutrients on the back of the latest restrictive health craze.

13

Artificial Sweeteners and Preservatives

John (Long John) Daly was a lucky guy. It was 1991 and he had just got his professional golf association card and was sitting at home in Arkansas when the phone rang. He was on a waiting list of nine others to see if anyone dropped out of the PGA tournament. It turned out that a golfer's wife had delivered a baby early and he rushed home. The other eight hopefuls on the list above him couldn't make it in time. He flew to the course, borrowed a caddy and without even playing a practice round found himself ahead of the field. Against all the odds he eventually won the tournament. The rookie was an instant crowd favourite for his lack of etiquette, big hitting and carefree attitude. He went on to win the British open in similar style. But then his inner demons took over and he stopped winning.

He began overeating, he couldn't control his alcohol intake and his performances suffered. He was also addicted to cigarettes (forty a day) and lost millions of dollars of his earnings on gambling. In his forties, after his coach left him with the parting line 'The most important thing in your life is alcohol', he went into rehab. He gave up alcohol and replaced it with Coke, then put on weight and changed to Diet Coke, then quickly became addicted to that as well, but his weight stayed the same.

A year later he had a gastric-band operation so as to lose weight. But, he complained, 'The band won't allow me to drink as many. If I don't have ice, I can't drink it. I can't have it straight because of the carbonation. I used to have twenty-six to twenty-eight cans a day. Now I have ten to twelve at the most.' He still plays golf and attends the US Masters, but only to sell his merchandise from a bus next to his office – the Hooters' Diner, known for its scantily clad young waitresses.

'Diet' drink addictions like John Daly's, or for the 'real thing' with sugar, are increasingly common and certainly a cheaper habit than alcohol or cocaine, but they have disastrous results on the body's metabolism. Clearly, Daly's problems didn't go away when he changed from regular to diet drinks. Most cases like this don't qualify as a full-blown chemical addiction as he didn't have documented withdrawal symptoms. He just had the powerful urge to keep drinking more.

Artificial sweeteners – not the free lunch we once thought

Diet drinks appeared to be the ultimate modern invention when Diet Pepsi was launched, first in the US in 1963 and twenty years later in the UK. Not just the idea but the consumption of non-caloric sweeteners in foods has been around for a hundred years. An increasing number of people particularly turn to diet drinks with near zero calories to try to avoid the effects of sugar and reduce weight.

Not everyone likes diet drinks. Some people with over-sensitive taste buds and certain gene variants find the artificial taste strong and unpleasant. Others dislike the aftertaste. Some of the distaste is also due to the fact that we are very sensitive to the different textures and structures of the chemicals in the drink that are trying to mimic the taste of the sugary equivalent. The carbonation is another factor that can trick the brain into thinking a product is less sweet than it really is.[1] Flat cola is often undrinkable.

Since the 1980s diet drink sales have steadily increased globally. By 2014 they accounted for a third of total US sales and a $76 billion market. But the saccharine tide is turning. Growing public concerns about the health effects of the sweetener chemicals – particularly fears about cancer – have seen US sales drop since 2010, and European markets are likely to follow as people switch to caffeinated energy drinks. Most people, however, believe non-caloric artificial sweeteners are good for losing weight. Short-term studies in which overweight kids are switched from regularly drinking sugary drinks to artificial sweeteners are usually reported as being beneficial for weight loss. If you look closely, though, you see the results are not as clear-cut as you might expect, considering the substantial calorie differences.

The largest of these studies so far was performed on 641 Dutch children randomised to drinking a can of either diet or regular cola a day for eighteen months.[2] Both groups continued to gain weight over time. The diet cola group certainly gained less weight than the regular cola group, but not dramatically less, and the average weight gain of the diet group was disappointingly more than expected, with only minimal differences in feelings of fullness compared to the sugar group.[3]

A number of observational but longer-term studies have shown associations in the sweetener group with weight gain and diabetes, even after controlling for the fact that heavier people are more likely to use sweeteners in the first place.[4] [5] This could partly be explained as long-term psychological effects that alter behaviour. In another study 114 students were randomised to drink regular Sprite (a sugary lemonade), Sprite Zero containing aspartame, and sparkling water as a control. They showed that the diet drink altered some aspects of the students' future behaviour that made them seek extra calories and suggested the chemical had effects on the brain.[6]

This may not be so crazy. Aspartame, a key ingredient of the world's most used sweetener, can affect the brain cells of the hypothalamus and in theory upset appetite pathways.[7] [8] Other studies have shown that the reward pathways in the brains of habitual diet drinkers have altered so that they get a greater kick out of sugar.[9] These chemical molecules that tickle and trick our taste receptors out of calories are now in many of the foods and drinks we consume and hard to completely avoid.

The world's most flexible sweetener used in food, soft drinks and alcohol has the friendly name sucralose, but it is actually the less catchy 1,6-dichloro-1,6-dideoxy-β-D-fructofuranosyl-4-chloro-4-deoxy-α-D-galactopyranoside. It used to be thought of as an inert chemical possessing five hundred times the sweetness of sugar that passed straight through the body and sailed through safety tests for cancer. But although there is no good evidence – as opposed to public opinion – for linking its use with cancer, as usual there is more to the story.

Several studies have now shown that 'inert' sucralose alters hormones involved in digestion. It does this by activating the taste receptors, which as well as being on the tongue are also found on the pancreas, in the gut and in the hypothalamus. Small human studies of obese patients show this leads to increased insulin release, increased gastric emptying and the release of normal digestive hormones such as GLP-1.[10] Sucralose also alters liver enzymes in rodents that in humans can modify how we respond to a common array of drugs.

Sweeteners are clearly not 'inert'. Digestively they arrive largely untouched in the colon, where they can interact with our microbes. There appears to be a good degree of individual differences in how we respond to sweeteners, which could be down to our microbe communities. Some evidence of this came from early studies in 2008 looking at just a few microbes in rats. Researchers fed rats Splenda (sucralose) at the FDA-recommended human doses for twelve weeks, and found significant reductions in total microbe counts and diversity, and in particular this affected the healthy microbes.[11] The gut was also made more acidic. Some of these changes were sustained for up to three months after the drinks stopped. An observational but detailed diet study of 98 subjects, although not designed for the purpose, suggested for the first time that there could be an association between aspartame intakes and changes in microbe contents in humans.[12]

Alkaline diets and our microbes

Interest in trying to change the acidity of the gut via diet has waxed and waned in the world of nutrition, but was rekindled when Alkaline diets became popular. The theory is that if you reduce acidic foods you can make the gut less acidic and the blood more alkaline and 'healthy'. As before, regarding other explanations involving the body's acids, the theory is rubbish. The gut is naturally designed to be very acidic to digest food and the blood is kept slightly alkaline. The body controls the blood acidity very tightly via your kidneys and urine, so diets can't affect it. But as many of the constituents of the Alkaline diet are vegetables and it involves avoiding meat, this diet myth may have some inadvertent benefits.

Although alkaline foods alone cannot change the acidity of the gut significantly, some common drugs can. Anti-acid-secreting drugs called proton pump inhibitors (PPIs) are one of the bestselling drugs in the world, and can usually be bought without a prescription. They are very good at easing symptoms of heartburn and ulcers and one in seven people have used them. They are popular as they are thought to have few, if any, side effects. However, we showed in 2016 that in 1,800 twins they caused major shifts in their gut microbes, with big increases in microbes that normally live in the upper part of the intestines (like streptococcus). This could explain why people taking these drugs regularly have increased rates of gut infections and possibly other adverse effects, such as slight increases in fractures or cancer.[13] It is also another example of how drug safety tests have overlooked the impact on our microbes.

One of my PhD students, Madison, 'volunteered' for a short study to monitor her gut microbes as they received first a Coke then a Diet Coke. The plan was to get baseline results, then have three days of the real thing, sugar Coke, drinking 1.5 litres a day of it, followed by three days of Diet Coke, then just water for three days. Analysing one person is problematic, and some of her samples had to be redone because the sequencing, as in real life, doesn't always work perfectly. Madison was happy when we looked at the results, as they showed very high levels of the rare (and tough to pronounce) microbe family Christensenellaceae that we found to be protective for obesity in our twins study. But she shouldn't have worried because she is fit and lean anyway. However, we also found surprising changes on the Diet Coke days, which showed an increase in the diet-responsive microbe Bacteroidetes that was hard to interpret.

Try taking antibiotics with your diet drink

Luckily, at around the same time an Israeli group led by Eran Elinav and Eran Segal had a similar idea but did the study on a much grander scale and published their findings in the journal *Nature*. First, they found that giving normal supplement doses of three common sweeteners (sucralose, aspartame or saccharine) to mice on normal or fat diets produced a significant increase in blood-glucose levels compared to the sugar-consuming mice. They repeated the test with antibiotics to kill off any effect of microbes, and this completely removed any effect of the sweetener.

Next, by transplanting them into the germ-free mice and producing the same glucose increase in the new animals, the researchers showed that the gut microbes were directly responsible. They then looked at the microbes of forty people participating in a nutrition study who regularly consumed sweeteners compared to 236 who didn't. They found exactly the same effects as in the mice: namely, abnormal blood-glucose and insulin levels. They then set up another study in which seven sweetener virgins were given saccharine supplements (at normal approved levels) for seven days on top of a standard controlled diet while having their blood glucose monitored.

Although people responded differently, four out of seven experienced major changes, including increases in Bacteroidetes and quite a few unusual gut microbes, which mirrored their glucose changes. The sweeteners induced the microbes to overproduce two of the metabolic signallers we mentioned previously, the short-chain fatty acids (SCFAs), but unusually these did not include the healthy butyrate. Overall, the function of the new microbial team was ramped up by the sweeteners to digest carbs and starch more efficiently than before. This would affect their digestion of normal food, which in turn could explain the weight gain.

This elegant series of experiments shows that artificial sweeteners are definitely not a free lunch; they do have potentially harmful metabolic effects that can increase weight gain and the risk of diabetes. They do this because even so-called inert chemicals can be crucial for our microbes, which change their function and so affect our bodies. We don't yet know the true extent of the risk of the sweeteners and whether everyone is susceptible, but these microbe experiments have made sure that we and our food regulators who approve the new 'safe compounds', once they pass the cancer tests, should now take these risks more seriously.

Manufacturers of diet drinks and many processed foods, because they lack the natural anti-bacterial effect of sugar, compensate by adding large numbers of chemical preservatives such as sodium or potassium benzoate, citric acid or phosphoric acid. Many of these, including the benzoates, tartrazine, MSG, nitrites and nitrates, are commonly reported as causes of allergy. The same chemicals are also likely to be having major unsuspected influences on our microbes by potentially reducing their numbers and their diversity. They also affect our immune systems, both directly and via interactions with our microbes.[14] This has been suggested as another reason for the recent rise of allergies. Most of the chemical safety-testing of food additives and sweeteners focuses on determining the risks of poisoning or cancer, not on detecting metabolic changes in our bodies. So until we know more, maybe we should be reducing or, ideally, avoiding these 'harmless' chemicals.

The global rise of chemical sweeteners in drinks and foods looks as if it is slowing down as some of the public turn back to natural products. But as noted earlier, the downside is that this usually means using more sugar. The soft-drink companies are trying to keep up with health-conscious customers by using – since 2011 in the EU – the sweet stevia leaf as a 'natural' alternative to sweeteners. It is claimed to reduce calories by 30 per cent. Because of its cost it is usually mixed with cheap sugar. It allegedly has no downsides, but it hasn't been tested properly. By the time you drink this latest miracle sweetener in a manufactured drink, it has been heavily chemically processed; some people, though, can still detect its aniseed-like taste.

But what about natural stimulants that have been used for millennia – are they as bad?

14

Contains Cocoa and Caffeine

The very first article I wrote and published while a medical student showed international data linking coffee consumption with cancer of the pancreas. This is a nasty and usually fatal cancer that was increasing at the time in the West, and the data I found suggested coffee could be a cause. This paper was a real breakthrough, not for science, but for my CV and my research career, which took off as a result. As I said before, such ecological studies are usually flawed. The same analysis would have shown that owning a TV or flared trousers, the incidence of which had also increased in the same regions, was associated with the cancer, and the media headline would have been that it could have 'caused' it. And the same analysis would have shown that chocolate (cocoa) eating was deadly.

Anyway, another thirty years of similar bad science and media scare stories followed, along with hundreds of expensive research studies using observational data and overdosing poor old rats with coffee. A similar story emerged in the 1980s telling us that hormone replacement treatment was thought to be a miracle cure for everything from depression and heart disease to dementia and declining sex drive. I was one of those taken in. Only when clinical trials were performed did the risks of heart disease and cancer emerge and the benefits disappear, with the exception of preventing fractures. We have learnt lessons from past mistakes, but it often takes time to reverse old ideas and prejudices. The jury is now in as to whether coffee and its accomplice caffeine are guilty as charged, and whether chocolate and cocoa are really naughty but nice.

When our mid-year exams during our first year of medical school arrived, my friends and I were panicking one day about a tough biochemistry exam that we hadn't prepared for. We decided we all needed to revise all night to stand a chance. We took caffeine (Pro Plus) tablets, which I knew would certainly keep me awake and alert and I was confident of a solid twelve hours of swotting before the exam. The reality was somewhat different. I was trembling and twitching and couldn't concentrate or remember anything. Exhausted and lacking any facts worth mentioning, I failed miserably. I was wary of caffeine for many years after, and like most people believed that it and the other famous bean extract, cocoa, were bad for you.

Caffeine is probably the world's commonest psychoactive drug, drunk by 80 per cent of the world. But it is addictive in many coffee drinkers, who develop shakes, attention problems and withdrawal headaches if they stop suddenly. The effects of complex foods like cocoa and coffee can now be studied more precisely. Our bodies contain thousands of circulating chemicals, metabolites, whose importance is reflected in how our cells function metabolically. We can now measure most of these individual chemical fingerprints quite precisely in just a drop of blood, saliva or urine with the science known as metabolomics. We have used this technique in our twins, with spectacular gene discoveries and new links to diseases, and of the 1,200 existing metabolite signatures that we can now identify in blood at least 250 are produced solely by our gut microbes.[1]

Several studies in humans have shown that microbes, probiotics and antibiotics can alter the levels of the key chemicals that are the precursors of brain neurotransmitters. These chemical signals, such as tryptophan and serotonin, play a pivotal role in the brain as well as in depression and anxiety. Most of the body's hormone serotonin is produced in the intestines, and recently we have found it is mainly manufactured by microbes during times when we are not eating. Autism is increasingly and consistently associated with disordered microbes, and these could be the link with abnormal brain-chemical signalling.[2]

Is chocolate really a miracle food?

Studies showing the miraculous powers of chocolate are much loved by the media and the public, especially in the UK. We are third in the chocolate-eating league tables at 9.5 kg per person in 2012, behind only Ireland and Switzerland, and nearly double the US. The feel-good effects of chocolate are in part due to our gut microbes.

In several studies, dark chocolate given to adults produced significant changes in the blood neurotransmitters and other metabolites that only microbes could have produced.[3] There is also evidence that obese people are more sensitive to the alluring smell of chocolate than skinny ones.[4] So if you like chocolate a bit too much, blame it on your microbes, who like to keep your brain happy.

Unlike with many other easy-target foods and additives, such as e-numbers, colorants, fats and burgers, the 'chocolate is healthy' stories are regular features in our newspapers. For instance, it is claimed to miraculously prevent or cure heart disease, cancer, depression, low libido and sexual dysfunction. The source of these acclaimed 'special powers' is a bean that is roasted and fermented. This is the fruit of the Theobroma cacao plant, literally 'the fruit of the gods' and supposedly first cultivated by the Aztecs. Most chocolate we eat is made up of sugar, fat, milk solids and cocoa beans. Culture and marketing determines the amount of milk and the percentage of cocoa in the mix.

There is considerable worldwide variation, but Anglo-Saxon countries generally prefer the milk variety, although dark chocolate consumption is increasing. While, in our twins, liking sweets and chocolate in general has a genetic basis, we found the preference for milk chocolate over dark was mainly cultural with little effect of genes. Within chocolate eaters, genes rather than culture played a greater role in whether you prefer hard or soft centres. This is likely due to genetic preferences for different textures as well as for sweetness.

Regardless of the media hype, the absence of long-term studies and my initial scepticism, the evidence is now pretty good, though not water-tight, that the cocoa in chocolate, which itself is made up of over three hundred chemical substances, is positive for reducing risk factors for the heart.[5] A few, admittedly observational, studies have even suggested that regular consumption is associated with lower body weight.[6] Over seventy human clinical studies and many more animal studies have been performed on cocoa.

Of the many chemicals in the substance, the constituents showing the clearest benefit are compounds known as flavonoids. These are part of the same polyphenol family found in nuts and olives that we discussed earlier, which has anti-inflammatory, antioxidant and important microbial effects.[7] Gram for gram, cocoa has the highest concentration of polyphenols and flavonoids of any food, and it's a precious commodity.

Chocolate eaters are often embarrassed to be questioned on the subject, and as with calorie and alcohol intakes, people are often 'economical with the truth' about their consumption. So as to avoid this, with colleagues in Norwich we looked at metabolomic markers in the blood of 2,000 of our UK twins. We found that twins with the highest blood-flavonoid levels coming from chocolate, berries and wine had lower weights, better arteries, lower blood pressure, stronger bones and a lower risk of diabetes.[8] This all seems too good to be true – and it *was* observational – but it makes more sense when we consider the role of our microbes. In fact, there is now compelling evidence that our microbes, like us, enjoy chocolate. In the gut they play a major role in the metabolism of the chemicals from cocoa that lead to improvements in blood-lipid levels. When a British clinical trial gave volunteers the polyphenol extracts (flavonoids) from cocoa for four weeks, it saw significant increases in bifidos and lactobacilli and reductions in the Firmicutes family as well as in markers of inflammation.[9]

The authors of the trial report suggested these cocoa flavonoids could make a useful prebiotic supplement and the Web offers plenty of ways to buy them. High-grade cocoa can now be bought in health food shops, and confectionery giants like Mars are now marketing 250 mg cocoa flavonol supplements such as Cocoavia™ as powders to add to milk, oatmeal or smoothies.[10] The downside is the extra 200 calories you need to ingest to get the benefits. The EU has approved a health claim from a Swiss chocolate company for a similar product, and many more will follow.

Less is known, however, about what happens in the real-life situation when you combine the cocoa with the sugar and saturated fat that you find in a normal chocolate bar, which most people regard as unhealthy. This was recently carefully tested in Swiss volunteers using 25 grams of a 70 per cent cocoa dark chocolate, twice a day for two weeks.[11] Each serving (Nestlé Intense, since the study was sponsored by them) comprised also 6 grams of sugar and 11 grams of fat. The subjects ended up with no worsening of their total or bad

(LDL) cholesterol and there was a significant increase in their good cholesterol (HDL), despite the extra fat. Another Swiss study of dark chocolate eaten for four weeks found improvements in blood vessels, but adding extra flavonols to the chocolate made no difference.[12] Other studies showed similar beneficial short-term effects on lipids.

The microbes fed off the flavonoid polyphenols and produced many helpful by-products, like the healthy SCFA butyrate. What was striking was that after a week the amount of these polyphenol metabolites was greater than could be accounted for just from the diet. The microbes, once fed a bit of chocolate, were producing these healthy chemicals of their own accord, like little home industries. They also found that regular chocolate eaters had different and healthier metabolisms and microbes than occasional chocolate eaters.[13]

Beware the milky variety

So that's fine if you like 70 per cent cocoa dark chocolate, but what if like most children, Brits and Americans you prefer your chocolate milkier, like Cadbury's or Hershey's? Is this kind helpful or harmful to your microbes and health? The UK market leader Cadbury's Milk was launched in 1905 as the 'first' real milk chocolate (although this was likely invented in Germany in 1839). It now contains 26 per cent cocoa, plus cocoa butter, which is the mainly saturated fat coming naturally from the cocoa nut, cow's milk and sugar. A small serving of four squares contains 4.7 grams of saturated fat and 14.2 grams of sugar (3½ spoonfuls). Cadbury's milk chocolate used to contain 23 per cent cocoa, then increased it in 2013 because EU milk chocolate now has to contain real milk and at least 25 per cent cocoa. In the US you only need a miserly 10 per cent cocoa to call it chocolate. Hershey's milk chocolate is only about 11 per cent cocoa and proportionally sweeter. Each serving contains 8 grams of fat and 24 grams of sugar (6 spoonfuls). Although milk chocolate uses the same type of cocoa as dark chocolate, to get the polyphenol benefits you need to eat three to five times more, and you'd likely get fat and lose your teeth in the process.

So training your taste buds to eat more than 70 per cent cocoa dark chocolate may be worthwhile in the long term. But working out your minimum daily dose is going to be difficult, as actual amounts of helpful flavonoids vary widely according to the manufacturing process. Some companies sneakily make the chocolate darker with additives and the percentage of cocoa stated is only a rough guide. Hopefully, flavonoids and polyphenols will soon be included on the food label.[14]

Food company scientists noticed decades ago that combining lots of fat and sugar together can produce near-addictive qualities. Add in the magical properties of chocolate, and it becomes irresistible. Most of the bestselling confectioneries use this formula and companies are continually trying to improve their blends so as to tease the consumer's taste buds. One of the latest potential blockbusters was launched by Kraft foods after it merged with Cadbury's to become the global leader. It is called Kraft Indulgence Spread, a mix of Kraft Philadelphia Cream Cheese with 'real' Belgian chocolate. It may be popular, but don't expect highly processed foods like this to be microbe, polyphenol or waistline friendly. Returning to the Mediterranean paradox: residents of most of these countries – France, Spain, Italy – traditionally prefer strong dark (high-cocoa) chocolate to the milky and less healthy variety. This preference might well contribute to their better cardiac health, although it would be hard to prove.

Caffeine can let you down

Michael Bedford was at a party near his home in Mansfield, Nottinghamshire. He ate two spoonfuls of caffeine powder he'd legally bought online, and washed them down with an energy drink. He began slurring his words, then vomited, collapsed and died. He had ingested more than 5 grams of caffeine, equivalent to about fifty espressos. The coroner cited caffeine's 'cardiotoxic effects' as the cause of death. He was twenty-three years old. Caffeine is not always harmless fun.

The soft drinks companies add caffeine routinely to enhance taste, complexity and bitterness and give the drinks a kick, especially in diet colas lacking real sugar. Levels in a can of Pepsi and Coca-Cola have reduced in recent years, as some parents complained of overexcited insomniac kids. Levels vary between the US and Europe, with higher levels in the US. Pepsi Max contains 43 milligrams in the UK and 69 in the US; Coke, 32 mg here and 34 mg in the US; Diet Coke, 42 mg and 45 mg. Other drinks companies like Red Bull (80 mg) make caffeine a part of their appeal to all-night-partiers needing 'energy'. Most regular fizzy drinks average around 30–45 mg, about half that of a normal coffee or the same as a mug of weak tea.

Companies add caffeine to most drinks officially for flavour and complexity, but unofficially to improve their addictive properties on top of the already powerful fructose. Tea contains variable amounts of caffeine (20–70 mg) but usually has half that of coffee per serving, and strength is unrelated to colour. Unsurprisingly, among the UK media, tea always had great PR and was blessed with medicinal properties. The publicity seemed to be backed up with reasonable epidemiological data from a summary of twenty-two prospective studies showing a 25 per cent reduction in mortality for those drinking two to four cups a day, dropping off above that amount.[15] The caveat is that benefits seem to be only for green-tea drinkers rather than black-tea lovers, as in the UK.

The unpredictable espresso hit

In the UK coffee has replaced tea as the commonest beverage, and the success of the coffee franchises around the world suggests this is part of a global addiction. Caffeine and its main beverage coffee have come under a huge amount of scrutiny and suspicion and have been the target of many media scare stories. Caffeine has been linked with increasing stress, poor sleep and long-term heart disease and cancer. It is very hard for people to know how much caffeine they are drinking with each cup of coffee as there are no standards. A UK study of twenty espresso bars found a nearly fourfold difference in the caffeine in a single espresso shot, which could be as high as 200 milligrams, the recommended upper limit for pregnant women. There was found to be even a large daily variability in samples from the same coffee bar, and between countries a tenfold variation in a single shot.

The lack of a standard cup of coffee explains the unpredictable effects, and an Italian can still sometimes be surprised by the strength of an American brew.[16]

It turns out, though, that drinking even six cups of coffee per day overall is not harmful for most people. Combining twenty-one individual prospective studies of the coffee habits of over a million people from Europe, the US and Japan that were followed for several years until 128,000 of them had died, provided powerful data. It showed that moderate coffee drinking, three or four cups per day, reduces the risk of death by around 8 per cent and heart disease by 20 per cent. Unlike my own, inferior, analysis, there was no effect either way on cancer.[17]

Although we should still be wary of epidemiological data because other linked habits of coffee drinkers (such as drinking alcohol or smoking) could be responsible for the findings, since a long-term clinical trial forcing people to drink coffee is an unlikely eventuality these estimates are as good as we will get. A few people with sensitive hearts should avoid extra caffeine, but for most of the two billion people who drink it daily it appears beneficial.

So if coffee really is good for us, why might this be? It is addictive and in excess can cause anxiety, poor sleep and heart arrhythmias. In the UK we are still wary of it and mainly because of the caffeine content.

The cocoa plant evolved to produce caffeine in large amounts mainly to fend off predators nibbling at its leaves, where the highest concentrations are found, and to give it room to spread out in the surrounding soil. It may be that some susceptible animals became addicted to the caffeine, kept returning for another nibble, and spread the seeds around. Geneticists have recently found that the cocoa plant and humans share an important characteristic: their genes. Cocoa actually has more protein-coding genes (25,000) than humans (around 20,000), which should improve our view of the evolution of coffee – or lower our view of humans.[18]

Scare stories still come out about potential carcinogens in different types of coffee and in the bean-roasting process.[19] Caffeine is just one of thousands of chemicals the beans contain, and numbers increase further on roasting. Most internet accounts misleadingly focus on the properties of just one single ingredient out of the thousands. Arabica coffee contains at least twelve different polyphenol compounds, the commonest of which has a nasty name conceivably more appropriate to swimming pools – chlorogenic acid. However, so far we think that most of the polyphenols, like those in cocoa and olives, even if they sound toxic are actually beneficial to us.

We performed our twin studies to show that preference for coffee, as for many foods, is under strong genetic influence.[20] This may be partly to do with taste genes, but collaborating with another team in a gene-association meta-analysis we found these preferences depended on different specific genes which control the way coffee and its thousands of chemicals are processed by the liver. The good or bad feelings we get from complex foods like coffee are also dependent on the chemicals that our bodies produce when our enzymes break down the food product; in other words, it could affect people differently.[21]

As well as the thousands of chemicals and the dozen antioxidant polyphenols, coffee has the surprisingly high fibre content of half a gram per cup. This combination of fibre and polyphenols permits it also to provide food for our gut microbes.[22] Microbes break down the fibre to produce the key short-chain fatty acids like butyrate, which allow other helpful Bacteroidetes and Prevotella species to increase.[23] So coffee wakes our microbes up in the mornings, and it's not just the caffeine that does it.

What happens if you don't like espressos or strong brews? In the US you may need the jumbo pint-sized portion to get enough polyphenols, but our microbes may be nearly as happy drinking the polyphenols in decaffeinated or freeze-dried instant varieties if the doses are equivalent. The same may not be true for tea lovers. Although tea (whether green or black) contains some polyphenols, it lacks the extra fibre of coffee; small differences in the oxidation and fermenting process could explain the observed health benefits of green over black tea despite them coming from the same leaf. For those who don't like to wake up to the smell of actual coffee, there is some evidence, if scanty, of a potential for weight loss if you use green coffee bean extract as a supplement.[24]

The stories about foods and drinks such as coffee that transform them from villain into hero and vice versa should make us all more wary of accepting at face value the evidence of any common food as deadly or as a miracle cure. Nevertheless, it is comforting that our current knowledge strongly suggests that our habit over thousands of years of consuming cocoa and coffee may not be harmful, and may even be a good thing. But can the same be said of the demon drink?

15

Contains Alcohol

The yellow death started at the end of 2006, when hundreds and then thousands of people started turning up at Russia's A&E departments with a vivid yellow skin and yellowing whites of the eyes. They were very sick, with vomiting and itchiness. Some died within a week and others survived for a few years if they were lucky. Natasha was a thirty-year-old from just outside Moscow, a single mother with a seven-year-old boy, who would be dead from liver failure within a year. All the yellow patients had one thing in common: they had drunk moonshine (*samogon*, in Russian), which was in reality a cheap mix made from 95 per cent ethanol medical antiseptic that they had bought for 40 pence a bottle. In total, 12,500 people were affected and about 1 in 10 died in that outbreak alone. The locals were not sympathetic to these alcoholics.

The Russians have been some of the heaviest drinkers (and smokers) in the world for over sixty years. Consumption peaked in the early nineties after the disintegration of the USSR, when alcohol death rates temporarily reached 40 per cent. Today an estimated 25 per cent of Russians has an alcohol problem. Roughly half a million die every year, and alcohol kills 1 in 4 men before they are fifty-five. This gives them the lowest life expectancy in Europe, at sixty-four years for males, and puts them among the bottom fifty countries in the world. This excess mortality is mainly due to heart disease and cancer but is related to the average amount drunk per person which is 15.7 litres, double that of the US, but also to the binge drinking of vodka.[1]

Is alcohol good or bad for you? The answer is not clear-cut as it depends how much you drink, and who and where in the world you are. It is clearly bad for Russians, but what about Mediterranean populations, for whom it is part and parcel of diet and lifestyle?

We are told many conflicting things. Alcohol can be poisonous and addictive, can cause malformations in babies, lead to cancer and depression; or alternatively it can enhance mood and social and sexual success, alleviate heart disease and promote longevity. We are told that safe upper limits for men and women vary between two and three units (for instance, glasses of wine) a day. But these are guesstimates based on averages, and everyone's glass is different. In Europe and many other countries, alcohol is exempt from the usual requirements of food labelling, so the contents are obscure. The calorie count of 180 for a pint of beer and 150 for a glass of wine surprises most people. In many countries and cultures beer is not considered to be alcohol, which only increases the confusion.

But it's not surprising if we struggle with alcohol, given that we still have problems with something as simple as water. Most of us believe that we need more than two litres (eight glasses) of plain water a day to stay healthy. Many young people nowadays go about with plastic water bottles – full of chemicals and devoid of microbes – attached to their person, for fear of dehydration. This is another modern myth for which there is no supporting data. We vary greatly in how much water we need, and our bodies are perfectly adapted to tell us when we are thirsty. Anyway, we get much of our water from food, coffee, tea, soft drinks and even alcohol. Our hunting ancestors didn't have to swig from bottles every five minutes to survive.

Blame it on the genes

It was 2.36 a.m. on a warm night in California in 2006, and cruising along happily in his Lexus on a Malibu coastal highway was the Hollywood star Mel Gibson. He was pulled over by a highway cop, James Mee, as he was speeding at 84 m.p.h in a 45 m.p.h zone. His travel companion on the passenger seat was a nearly full bottle of tequila, which he admitted he had had a few slugs from. He was co-operative until the breathalyser showed him to be 50 per cent above the legal limit and the officer arrested him. Then his inner demons took over. According to the police report, Gibson became angry and aggressive when not allowed to drive home. He at first despaired: 'My life is over. I'm f****d. Robyn's going to leave me.' Then he bragged that he 'owned Malibu' and would 'get even' with the officer, then asked him if he was Jewish and said: 'F*****g Jews, the Jews are responsible for all the wars in the world.' When the report was posted on the internet, the worldwide public and media backlash was enormous, and is still talked about today. That same day his wife of twenty-six years separated from him.

The actor apologised profusely, saying that he had been under a lot of pressure and had a lot of pent-up anger, and that 'everyone does this kind of thing'. He might also have tried blaming his genes. He comes from an ultra-conservative Catholic family, and his father a few years before had caused a similar uproar by making anti-Semitic statements and claiming that the Holocaust was 'mostly fiction'.

We know from previous twins studies that those very human characteristics of aggressive behaviour, extreme conservative views and even religious beliefs have a significant (around 50 per cent) genetic influence.[2] But this gene defence doesn't go down very well with judges as it suggests, perhaps unfairly, that people may never change. The question most people ask is whether alcohol just disinhibits people, allowing their true thoughts and beliefs to emerge, or whether a combination of alcohol and anger can trigger random insults and tirades. Gibson blamed the alcohol and his state of mind. He lost his licence for three months and was sent for alcohol remedial classes. If he had claimed that his problem of alcohol addiction ran in the family, which his brother Chris also suffered, would he have got more sympathy?

All the ways we use of dealing with alcohol are under some degree of genetic influence. Some people don't like the taste, such as the bitter flavour of beer, others quickly feel nauseous or headachy. Some can tolerate large amounts with no side effects, and tend to be the ones who get addicted and turn into alcoholics. The ability to drink alcohol pleasurably depends on a few key enzymes that vary around the world. Asians famously have low tolerance, as they carry a variant of the alcohol gene dehydrogenase, which metabolises it fifty times faster than in Europeans or Africans, who lack that variant.

Often this means that sitting down to drinks with Japanese friends or colleagues produces red faces and giggling after only one glass. Some argue that the American West was won by means of a mutation in these genes, as 'white man's firewater' was the downfall of many American Indians, who had migrated from Asia with their genes ten thousand years before. The current hypothesis is that, as with changes to the amylase gene altering our way of digesting starch, a genetic adaptation has occurred since the advent of farming. Alcohol is converted in the body by the enzyme ADH into acetaldehyde, which causes all the nasty side effects like flushing, headaches, vomiting, amnesia and disinhibition. ADH is made by a gene that now varies widely between people and countries.

In China, rice farming started on a grand scale around ten thousand years ago, and fermenting the rice with microbes soon produced rice wine and other alcohol. We think that while many folk drank the home brew regularly without any side effects, some became alcohol-tolerant or alcoholic. These inebriated Chinamen, and possibly women, may often have found themselves face down in the paddy field rather than looking after their children. Lying in the gutter in a pool of vomit, they may not have been seen as desirable mates. And the offspring of women who drank in pregnancy may have been less likely to survive. So the tolerant alcohol genes died out and the highly sensitive (drunk on half a glass of lager) alcohol-metabolising ones prospered, as these people never got hooked and had many more children that survived.[3]

From about six thousand years ago this initially rare genetic variant multiplied and took over Asia, and soon became the norm. Alcoholics exist in Asia but they are rare, and usually they have the non-mutated European genes. This may also account for the success of karaoke bars in Japan and much of Asia, where people need only the smell of alcohol to lose their inhibitions. The history and evolution of alcohol genes in China suggest that alcohol was not of overall benefit to the population. But the gene mutation didn't spread very far into European populations, only a small percentage of whom are now alcohol-intolerant.

So why didn't the mutated genes colonise Europeans? Did European women actually find drunken men more attractive, or was there something in the alcohol that gave regular drinkers some advantage that counteracted the obvious problems of addiction and alcoholism?

Study after study around the world observing large populations has shown an association between a lower risk of heart disease and mild alcohol drinking when compared to total abstainers.[4] A study summing up thirty-four observational studies found risk reductions of around 18 per cent below 4.8 units a day for men and 2.3 a day for women. The maximum protection against heart attacks and mortality was seen for daily intakes of just less than a unit of alcohol per day, equivalent to a small glass of wine.[5] But when you look at heavy drinkers who have much higher rates of heart disease, like the Russians, the risk starts to go in the other direction (this is referred to as a J-shaped curve). Remember that these are observational studies and thus could be flawed by biases caused by other factors associated with drinking. One of the few randomised studies of wine drinking published in 2015 studied 224 teetotal patients with diabetes who were given a standard Mediterranean diet with a daily glass of either red wine, white wine or mineral water for two years. The results were surprisingly clear: the red wine improved blood lipids and glucose measures compared to water, and white wine was intermediate.[6] The French had for several decades, until rates declined in both countries, half the incidence of heart and stroke deaths than Britons. As they drank a lot more alcohol in the form of wine, this was the most popular theory to explain the 'French paradox' we discussed earlier. But the French love of wine is also shown in their very high rates of liver cirrhosis and other alcohol-related cancers, which shows that the difference between protection and risk is a fine one.

Red wine supplements or *vin rouge*

In 1965 the average Frenchman or woman consumed nearly five bottles per week of *vin rouge*; in the 1970s a regular sight at breakfast time at autoroute service stations was French lorry drivers enjoying a glass of cognac. Now that both the rules and society have tightened up, wine drinking has steadily fallen to a quarter of previous levels, as have road deaths and to a lesser extent cirrhosis. The French are still near the top of the wine league table, drinking now just less than a bottle a week each and outmatched only by that hard-drinking all-male club that likes to dress up in robes and stockings, the Vatican. The French are now in danger of drinking more beer than wine, as young people are buying less and less of it. The same trends are being seen in other oenophile countries, like Italy and Spain.

Even if red wine definitely did prevent heart disease, its effects at best are mild and wouldn't explain this huge historical difference. Culture could play a part. The way you drink alcohol may be important: regular, relaxed, all-hours French-style drinking patterns are seemingly protective, as opposed to the episodic, down-in-one Friday night 'happy hour' binges of many Britons, which in total cost the country over £20 billion in lost productivity.[7] Nor is it clear from the studies that the protective effect of alcohol is restricted to red wine. Some show that all kinds of alcohol do the same job, and this includes beer, which the health-conscious Brits freely enjoy in large amounts – but, alas, without the expected health benefits if you exclude the old Guinness claims.

Belgium is known for its chocolate and artisanal beer, which really got going after the French Revolution when the fleeing monks wanted to drown their sorrows. You can buy over a thousand types of beer at one specialist Belgian café, most with their own distinctive glass and containing a huge range of ingredients including fruit. Belgian colleagues are busy working out the potential health benefits of the various beer polyphenols, but also believe that the yeast and prebiotics like inulin, with their infinite number of combinations and thousands of metabolites, can have positive effects on the gut microbes.[8] The beers range from super-strength 11 per cent alcohol ones that resemble wine, with healthy names like 'Delirium Tremens' and 'Mort Subite' (Sudden Death), to weak brews that up to the 1980s were served at school dinner times. In the current backlash against soft drinks in schools, maybe weak beer could have a renaissance.

Although the epidemiological evidence is unclear, and many wine drinkers also sometimes drink beer, there are other reasons to suspect that red wine may contain some chemical properties that could be healthy. The grape itself is extremely rich in polyphenols, which at the last count contained 109 different healthy chemicals including the subgroup flavonoids and the super-trendy compound resveratrol, which is profitably sold as a food supplement for non-drinkers.

Resveratrol is found in grapes, peanuts and some berries, has many effects on the body, and can regulate many genes epigenetically. If you believe all the studies and publicity, it offers a huge number of advantages such as increasing longevity and reducing heart disease, dementia and cancer. Twenty-odd years of intensive study have shown that taking large amounts of it, equivalent to over six bottles of wine a day, could reduce your cardiac risk, but only if you are a rat. Even after hundreds of rodent and test-tube studies there is still some confusion about the optimal doses.

Since we differ from rodents in the way we metabolise alcohol and resveratrol, we really need to perform studies on humans. Sadly, the data in humans has been disappointing both in quality and in results. One problem is that the compound is not readily absorbed by the gut into the blood, so unlike rats we can't get the necessary large amounts into the bloodstream. A few very small studies have suggested short-term benefits, but when the dose is increased to over 1 gram a day common side effects like diarrhoea occur.[9] And when doses are increased further, a few subjects experience toxic effects in the kidneys. So there is no evidence so far that it is good for humans.[10]

Do our gut microbes like a drink?

A detailed sequencing study of the gut microbes of ninety-eight American volunteers found that apart from the effect of body size and dietary fat, the one major dietary factor that had an effect on the composition of gut microbes was red wine drinking.[11] Recent data from around 8,000 subjects participating in the American and British Gut Projects run by my colleague Rob Knight have shown a big increase in microbial diversity in regular alcohol drinkers, which we believe is beneficial for our health, although we could not separate the effects of wine from Budweiser. So the association with our microbes could be with the alcohol itself or with the chemicals in the grapes or the beer. This distinction is not easy to grasp just by observational study.

One Spanish study, having managed to get ethical permission, found ten keen volunteers willing to drink alcohol regularly for money. They sampled three types of drinks over three months, taking daily two glasses of Spanish red Merlot, two of low-alcohol

(0.4 per cent) Merlot, and two shots of gin. The team examined the changes in their microbes, and the results showed increased microbial diversity, which we consider to be nearly always advantageous, in all three groups. The two wine groups (not the gin group) showed beneficial changes in certain key species. This was particularly seen as big increases in our old friends the bifidobacteria and Prevotella species, which also reduced lipids and signs of inflammation. This experiment suggests that it was the polyphenols in the wine rather than the alcohol that had the greatest effect.[12] It clearly couldn't differentiate between the possible benefits of resveratrol, the hundreds of other chemicals and the 109 polyphenols. Although it was often said that regular gin and tonics benefited the late Queen Mother's health and longevity, if she had added olives to the mix, providing her with yet more polyphenols, her microbes could have helped her to last even longer.[13]

Critics of the red-wine resveratrol health story say the amounts in red wine are too small to have large effects unless you drink about six bottles a day (which has well-known side effects). Also, it is mainly cheap red wine infected with small amounts of fungi that has the greatest resveratrol levels, and that is most likely to give you a headache. This is why resveratrol supplements are big business. Even big pharma got suckered into believing in the single miracle compound. The global giant GlaxoSmithKline purchased the Sirtris pharmaceutical company for $720 million. Their main product was a resveratrol claimed to act on the anti-ageing enzyme Sirt-1, which subsequently flopped.

When you start to think about how food interacts with our microbiota, it doesn't make sense to deconstruct it into individual chemicals. The power within it is likely to come from the totality of the ingredients of natural foods and the thousands of metabolic by-products observed when they interact with each other and with microbes.

Even if you, like me, cautiously believe in the possible health benefit of alcohol, you can have too much of a good thing. The average Italian and Irishman drink about the same amount of alcohol over a year, but the Irish appear to suffer more adverse health consequences. This could be genetic, but could also be due to a protective effect of the Mediterranean diet influencing the microbes. A few studies around the world examine the short-term effects of overdose on volunteers, usually hard-up students. But according to my friend Jaako Kaprio working at the Finnish Twin Registry, the amounts they are allowed by the ethics committees to give in a hospital, with full safety backup, are less than half the amounts taken regularly by Helsinki students on a good night out.

If you have ever suffered, the morning after, from an attempted overdose of resveratrol, you can now blame your microbes as well as the old excuse of food poisoning. A study in Kentucky gave volunteers a large single shot of vodka (140 millilitres) to mimic binge drinking, and observed what happened inside their guts and in their blood. Toxins released from the microbes' cell walls were rapidly seen in the bloodstream, at the same time as an increase in inflammatory microbes. These changes led to a stimulation of the immune system. The worse the effects of the vodka were on the volunteers, the worse the microbial disruption and the more microbe toxins were generated.[14] These LPS toxins from the cell walls have been shown in mice to activate the immune system and create an addiction to alcohol.[15] It may sound a bit far-fetched, but it's possible that microbes could be in part responsible for alcohol addiction in humans. But as always in studies of alcohol there was a wide range of responses, not just genetic but related to the differences in the species and diversity of the microbes already in the vodka volunteers' guts.

While large epidemiological longitudinal studies such as a recent one in the UK continue to report slight protective effects of moderate drinking, particularly in older women, there is still a worry about bias.[16] Several studies have used the genes for alcohol intolerance as a proxy for drinking rather than those unreliable questionnaires that rely on old-fashioned honesty. For years, from observational studies cancer of the colon has been associated with alcohol drinking, but a large gene-based study lacking most of the biases couldn't confirm this.[17] In 2014 a large meta-analysis study of 260,000 people using genes for alcohol intolerance (alcohol dehydrogenase) as a marker for drinking rather than using self-report questionnaires has cast doubt on the strength of the protective effects of light boozing. But the genes and their interaction with cultures haven't been fully sorted yet, and the studies didn't look specifically at wine drinking.[18]

So I'm hedging my bets and still drinking a glass of wine polyphenols – but not swigging bottles of vodka – and hope my microbes are enjoying them. Research in rats suggests that probiotics can prevent alcoholic side effects, so if you are planning a big night out, a glass of yoghurt may just help. Clearly, the latest research shows that what we drink and what effect it has on us is very personal. We all react in very different ways, due not only to our genes but also to our microbes, depending on the rest of our diet and our patterns of drinking. Consequently, blanket government guidelines on safe drinking (recently reduced in the UK to 14 units per week) will be too low for some and too high for others. Guidelines need to be tailored to us as individuals and, like tequila, taken with a pinch of salt.

In the UK more people take vitamins daily than alcohol – but are they much healthier?

16

Vitamins

Celebrities such as Rihanna and Madonna no longer pop pills. They reportedly now get their vitamins via intravenous infusions, personalised for them so that their bodies 'can absorb more vitamins and minerals'. Many people report a special energising vitamin buzz when they have regular top-ups in expensive spas or trendy drip-and-chill bars. Should we be following the celebs, or is this just an expensive hangover cure?

We all know that the foods we choose to eat are less healthy than fifty or even thirty years ago. Much of this is due to our love of processed foods, with virtually all of the original nutrients removed. What is less well known is that careful government surveys in the UK have shown that even our fresh fruits and vegetables and some meats now contain only half the amount of nutrients and vitamins as the same products did fifty years ago. Bombarded by advice and clever marketing, we have become preoccupied with vitamins as entities separate from their original food source. We take them for all kinds of reasons, most commonly to boost health, energy levels and ward off cancer.

The original vitamin (spelt 'vitamine' in the nineteenth century) was discovered from studying the Third World disease beriberi. Beriberi caused swelling of the arms and legs, heart failure and amnesia and other mental and nerve problems, and was of unknown cause. Then a Polish chemist, Casimir Funk, found it was due to the sudden change in people's diets from brown rice with husks to exclusively white polished rice. The rice had been denuded of its natural bran coating, which contained all the nutrients and the crucial vitamin B. He guessed many other diseases were due to a similar lack of 'vitamines'.

Many vitamins come from the food we eat. As well as synthesising a third of our own metabolites for us, our gut microbes also produce many vitamins. The B vitamins in particular, such as B6, B5, niacin, biotin and folate, and vitamin K, are manufactured by microbes for us. As mentioned earlier, vitamin B12 produced from microbes in the colon could be helpful for vegans, who don't eat meat, but is sadly not very useful, as it needs to be combined with hormones higher up in the stomach to be properly absorbed.

Some vitamins like folate and B12 can have an unpredictable effect on us, depending on the doses, by modifying our genes epigenetically (discussed in more detail in my last book, *Identically Different*). Vitamins such as vitamin A are crucial to our immune systems, and deficiencies have major impacts. Receptors in our guts, detecting the lack of a specific vitamin, change our microbes – and, in particular, reduce our protective long stringy (filamentous) bacteria, and this leads to immune reactions and inflammation.[1]

So keeping our vitamin levels normal is in both our own interests and our microbes'. Eating a balanced diet including regular fresh vegetables and fruit and the occasional piece of meat should maintain good levels for 99 per cent of us, but most people are not convinced. The first multivitamin supplement was produced commercially in the 1940s, and the market has grown steadily since. Around 35 per cent of Britons and 50 per cent of Americans take regular supplements, in a market worth £700 million in the UK and a staggering $30 billion in the US. Unless you are a shareholder in these companies, can this be a good thing?

In the past, our information on the benefits of vitamins has rested on Grandma's advice, on anecdotes, on what we've perceived to be the goodness of the original plant, or on extrapolating from serious cases of deficiency, like scurvy (caused by lack of vitamin C), or rickets (vitamin D). Then a few short-term observational and test-tube studies were done, which were not particularly convincing but certainly boosted sales, which continued to increase exponentially.

As we have heard before in other contexts, the idea spread that because fruit and vegetables appeared to protect against cancer and heart disease, then so should what the scientists thought were the crucial components, like carotene. In the 1990s several massive and carefully performed American epidemiological observational studies, following up health professionals, showed that taking antioxidant supplements like vitamin E was associated with reduced heart disease.[2] The world media reported this association as evidence of cause and effect, and everyone started buying antioxidants.

Oops – vitamin supplements cause cancer

In the mid-noughties some of these anecdotes, observational studies and marketing claims were finally tested in a number of properly randomised trials that looked at the trendiest antioxidant vitamins, especially carotene, selenium and vitamin E. They detected no benefit whatsoever for heart disease and in fact found a significantly increased cancer and heart failure risk in the groups taking them.[3] This caused a slight, albeit temporary, dent in sales. These observational epidemiological studies, as for many other diseases, had once again been very misleading on account of an inherent bias: people who chose to take vitamin E tablets were richer, better educated, thinner, drank less and were more likely to eat fruit and vegetables.

More recently, the benefits of multivitamins have been assessed and reported on with considerable attendant publicity. Reports included a meta-analysis of over 27 existing studies, and two new large randomised gold-standard studies of multivitamins, totalling close to half a million people. They have shown, convincingly, no benefits whatsoever.[4] The expert conclusions summing up all the collected evidence were damning. Beta-carotene, vitamin E, and high doses of vitamin A supplements are definitely harmful. Other antioxidants, folic acid and B vitamins, as well as multivitamin and mineral supplements, are ineffective for preventing mortality or morbidity due to major chronic diseases.[5]

Fish-oil capsules containing omega-3 compounds are widely marketed as a cure-all for the modern deficiencies in diet and lifestyle and for arthritis. Despite much hype and celebrity promotion, even when they contain the correct product they do nothing to help children's cognition, IQ or attention-deficit difficulties. Large studies of 12,000 high-risk cardiac patients have also reported their failure to reduce future heart disease risk.[6] As we noted earlier, the hype around changing omega-3 to omega-6 is misplaced, and other very large trials have now shown that fish oils fail to prevent macular degeneration, Alzheimer's and prostate cancer.

Vitamin C is the most commonly used vitamin in many countries. It is taken in the hope of boosting the immune system and reducing the risk of getting a cold. Proper trials, however, show it has no effect in the prevention of colds, or of cancer or any other disease. A few studies have shown that, like zinc supplements, it may reduce cold symptoms by half a day if taken early on, but an orange or some broccoli may do the same.

Most doctor-prescribed medicine is flushed down the toilet; less than 50 per cent is taken, even for serious conditions; and I see many patients who refuse all medications. But if you call the medicine 'a vitamin', some seem very eager to take them, even if they are known not to work. This is what I call vitamin loyalty. There is no proven benefit to the non-diseased person; in fact, the evidence so far suggests taking vitamin supplements regularly, especially in large amounts, has risks.

For example, overdose of folate is becoming very common, especially in countries like the US that already routinely supplement bread and other foods. Folate comes mainly from eating leafy green vegetables and fruit. Normally, we used to think, this was not a problem for adults and there was no upper limit. Studies (observational, of course) had suggested that extra folate could prevent heart disease and possibly cancer and increase fertility, and many experts were keen to add it to all foods or put it in the water supply. But genetic studies of folate in 2012 showed that the heart disease story was spurious. A dozen trials of folic acid supplements were started, to explore its anti-cancer potential, but a summary meta-analysis showed no benefit whatsoever.[7]

Pregnant or pre-conception women are a special case and advised in most countries to take around 2 to 5 milligrams daily of folate supplements, as studies have shown convincingly that treating all women reduces the rates of spina bifida (in which the spinal cord fails to close at twenty-seven weeks of pregnancy) and other birth defects. The benefits of supplementation are greatest in those with low starting levels, due to poor diets, in this case lack of fresh fruit and vegetables. The move was based on good science, and the incidence of birth defects has fallen as part of the successful public health campaign.

But what happens to women taking large amounts of additional supplements if their levels are already high? Or if, as often happens, they continue to take folates long term after the key twenty-seven weeks? Many people believe, wrongly, that the more you take of a vitamin the more advantageous it will be. Some anxious pregnant women take five to ten times the recommended dose of folate (just in case). Studies show that folate can epigenetically switch off some of the protective genes of both the mother and the baby, and in a high dose could be having other effects such as increasing the risk of allergy, asthma and breast cancer (apart from reducing the risk of leukaemia).[8] Other meta-analyses of randomised trials in 27,000 patients with heart disease have shown that folate supplements at 2 to 5 milligrams have no cardiac benefit, and have suggested that excess folate (over 5 mg a day) might increase the risk of heart vessel re-blockage in some people.[9] Other studies showed no benefits for infertility, and even a suggestion that it might increase the risk of it.[10] The mechanism is uncertain, and epigenetic changes to our genes could be a significant factor. In rat mothers fed large amounts of folate supplements, more health problems occur in their offspring, such as diabetes or altered neural brain connections.[11]

Broccoli sprouts or broccoli extract?

Most of the worrying folate evidence comes from rats who, as we know, are not always good models, or from small or observational human studies. Nevertheless, the possibility that overdosing on folate has health consequences is real and could extend to other vitamins. This also reminds us that the synthetic version of the vitamin, folic acid, may not have the same effects as taking in folate from, say, broccoli, which you can't overdose on. A 2012 study looked at this question specifically and showed in a clinical trial that administering either natural broccoli sprouts or equivalent doses of broccoli extract as capsules produced very different results. The natural product produced four times the amounts of the healthy polyphenols in the blood and urine compared to the artificial tablets.[12]

For over twenty years I was prescribing a gram of calcium plus vitamin D supplements to all my patients with osteoporosis, thinking I was helping them. This was based on some old studies and on the 'common sense' notion that calcium must be good for bone, as well as on the dogma that we should always give the same treatment as suggested by the trial results of more powerful bone drugs. I have been running an osteoporosis clinic for twenty-five years, and have never seen a pure African with a plain osteoporotic fracture. It always struck me as odd that most of the world, who don't drink milk and who take in a fraction of our Western dietary calcium, actually have fewer fractures, even without supplements.

As a result of the 1980s anti-fat campaign, many Westerners were advised to stop or reduce their consumption of dairy products. When patients mentioned this to their doctors, they resolved the problem by giving them extra calcium. In fact, it turns out that Europeans who regularly take calcium pills have stiffer arteries, full of calcium deposits. This gives them a slightly increased risk of heart disease and stroke which they were trying to avoid.[13] The area is still controversial with entrenched views on both sides, and clinicians are often the last to give up their habits of a lifetime, despite the potential risks and the lack of conclusive evidence showing benefits in fracture prevention. Nevertheless, if a clinician like me stopped a supplement and three months later the patient fell over and fractured herself, everyone would blame the dumb doctor. It is similar to the doctor's dilemma in withholding antibiotics.

But luckily, opinion and advice are changing, and routine supplementation with calcium and vitamin D for everyone is now being slowly discouraged and reserved for the elderly at risk.[14] Vitamin D is always in the news as it comes mainly from sunshine and to a lesser extent from diet in the form of oily fish, eggs, dairy and some mushrooms. Low levels of vitamin D are consistently associated by our old friend the observational study with an increased risk of over a dozen common diseases, including heart disease, high blood pressure, cancer, fibromyalgia, auto-immune disease and MS, as well as with depression and early death. With the possible exception of heart disease and MS, these claims are likely to be spurious, and I admit to having published some of those studies myself. Surveys of vitamin D levels across populations often report that a third are deficient. This has led to its being recommended as a supplement in a wide variety of situations as if it were a global panacea.

Sunshine not supplements

The sensible advice if you want to reverse a vitamin D deficiency is to sit in the sun for ten or fifteen minutes a day, exposing just your face and arms, or in winter to eat oily fish. But this advice is hardly ever given, and supplements are prescribed instead. This is due to our exaggerated fear of sunshine, perpetuated by cancer charities and suncream manufacturers. The advice is based on outdated observational epidemiological studies of skin melanomas. The public are told each spring that sunshine is *the* cause of melanoma. The reality is that these studies have shown that regular sunburning is associated with only a 50 per cent increase in melanoma risk. In other words, overexposure to the sun explains less than a quarter of cases at best.[15] Even this relatively modest risk fades when you adjust for people's skin type, pale or dark, which is determined by genes. In fact genes and bad luck are the main causes of melanoma, not sunshine.[16]

Nonetheless, most dermatologists tell their melanoma patients to avoid the sun at all costs. But it is low vitamin D levels and lack of sunshine that, paradoxically, make melanoma patients more likely to suffer recurrences.[17]

We have now placed our trust in doctors and vitamin tablets rather than in natural alternatives like oily fish or the sun, which are usually the better option. A while ago I ran a two-year randomised trial of vitamin D supplements in female twins with low normal levels, and found no differences between the bones of the twin taking the vitamin and those of the sister taking the placebo, although most other small studies that didn't adjust for genes reported an effect.[18] However, the truth was revealed recently when meta-analysis of over fifty trials involving over 95,000 patients taking vitamin D replacements showed no convincing evidence that supplements do reduce death or fracture.[19]

Three randomised trials using higher doses of vitamin D, at around 40,000 to 60,000 units a month (30,000 units is usually recommended), effectively rebrand vitamin D as a dangerous substance. The studies found that patients given higher doses, or whose blood levels were slightly above the average, had a third more fractures and falls compared to those on low doses or who failed to reach 'optimal blood levels'.[20] [21] Some of these low vitamin levels we are seeing in modern populations, rather than causing disease, may be markers for generally poor nutrition or a lack of outdoor activities, and this state of affairs may need to be put right first. Interestingly, a recent study of 100,000 Danes measuring genetic proxies of natural vitamin D, in an unbiased approach known as a Mendelian randomisation study, has shown that unlike artificial supplements which meta-analyses have shown to produce no benefits for cancer or mortality,[22] the right vitamin D controlling genes *can* (naturally) reduce mortality, especially cancer.[23]

There are a few rare exceptions to the general lack of evidence in favour of a benefit of using supplements in the absence of the obvious. For example, a combination of precise doses of two eye pigments, lutein and zeaxanthin, appears in long trials to prevent or delay blindness due to macular degeneration.[24] Getting the dose and ratio just right may be crucial in order not to cause harm.

Our bodies can cope with gradually extracting vitamins like calcium from normal food such as cheese, milk, broccoli or Italian mineral water, but they cannot cope with a sudden chemical surge happening in the stomach. Many hormones, such as parathyroid hormone on bone, can have bone stimulating effects on the body when given in large single doses artificially rather than the opposite effects when given slowly and naturally over a day. The same probably applies to many other vitamins not taken as nature intended, but artificially.

We know very little about the interactions between our gut microbes and vitamins, especially in mega-doses. Studies have shown substantial effects of vitamin B12 on microbe communities, and similar effects are likely with other vitamins.[25] Some of the side effects of synthetic or excess vitamins could be due to the mechanisms involved here.

It seems to me that the bottom line is this: unless you have a proven deficiency disease or are on a bizarre diet, vitamins don't help, and may cause harm to you and your microbes. Empty the bathroom cabinet of the whole family supply, and start afresh. We should be suspicious of the growing numbers of processed foods with 'added vitamins' until we have better data on their risks. Our obsession with the reductionist approach that isolates the one magic ingredient that can cure our ills is beautifully illustrated by our ill-fated love affair with vitamins. Focus on you and your children eating real food instead, which if reasonably varied contains most of the vitamins you need. Together with your healthy microbes you will manufacture the rest naturally.

17

Warning: May Contain Antibiotics

Antibiotics don't usually appear on the food label, but they should. We are all exposed to them even if we don't know it. This is one of the biggest changes to our environment in a million years, but has only had an impact during the last fifty. When the Scot Alexander Fleming, partly by chance, discovered in 1928 a mould that produced anti-bacterial chemicals he had no idea that our modern society would become so addicted to it. We now know it as penicillin.

In fact, Fleming never really saw the massive potential of converting the mould into a medicine. This was left to his colleagues Howard Florey and Ernst Chain, who carried out the essential process, then performed stunning trials on a few patients with normally lethal infections. The penicillin they made was so precious that they went so far as to collect each patient's urine so that after a bit of cleaning it could be reused for the next. Later, during the Blitz, they left wartime London and started industrial-scale production in the US. The aim was to use the drug for the Allied troops.

Antibiotics were an amazing success, saving millions of lives against the threat of often hitherto fatal bacterial infections. Doctors after the war predicted that antibiotics would herald the end of all human infections.

The DJ and TV-presenting twins Alana and Lisa (the Mac Twins) are twenty-six years old, successful and enjoying life. They are blonde, bouncy Scots, and being genetically identical look very similar. But scratch beneath the surface and they are more different than you might imagine. They are both the same height and similar weight, a healthy 9½ stone (60 kg), but Lisa has slightly bigger hips and once gained 2 stone in six months. Lisa now likes to control her weight with regular exercise, whereas Alana gets no kick from sports and prefers to sweat it off via 'hot [Bikram] yoga'. Alana finds it easy to do fasting 5:2 diets to control her weight, whereas her twin goes stir-crazy if she doesn't get regular calories.

Their personalities are also quite different. Alana used to be the shy one, and is much more pragmatic and steady, while Lisa sometimes gets anxious and is prone to bouts of obsessive-compulsive behaviour (OCD). Their father tragically dropped dead of a heart attack on the golf course aged only fifty-eight, and their grief reactions were nearly polar opposites – Alana stoic with episodic but dramatic breakdowns and Lisa in depressed denial. They had never worked out why they were so similar yet so different.

Raised in Scotland, they lived together in the same room for seventeen years and although they fought regularly were best friends. As six-month-old babies they were fragile, both had bronchitis that called for hospital treatment, then later recurrent ear infections and tonsillitis and consequently many courses of antibiotics. At the age of four, Alana had recurrent bladder infections necessitating several long stays in hospital and nearly constant antibiotics for two years. Soon afterwards she developed juvenile arthritis, a genetic auto-immune condition which affected many of her joints, causing painful swelling and stiffness. She was on many medications and managed to carry on a near-normal life, then at sixteen the pains suddenly stopped.

To the surprise of her doctors Lisa never developed any joint problems, but soon after leaving home she had a severe and unexpected late bout of acne. Alana never had acne, which is weird because our twins studies have shown it to be one of the most strongly heritable conditions. It was so severe that doctors put Lisa on several months of the antibiotic minocycline, followed by other, stronger, drugs to eventually clear it up. A year later, Lisa began to have urinary and kidney infections, which have continued to plague her, often calling for monthly courses of antibiotics. The doctors have even suggested she goes on them permanently.

Looking back, one possible reason why they are so different despite identical genes has to do with the antibiotics. Alana might never have developed arthritis if the natural gut microbes she received from her mother hadn't been decimated by the frequent antibiotics she took for her multiple childhood infections. These will have affected her immune system and maybe even caused her love of 'hot yoga'. Similarly, Lisa's late-onset acne, although having a genetic basis was caused by an overgrowth of microbes and an overreaction to them, while her later susceptibility to kidney infections also suggests a disordered gut microbiome as the cause. They both enjoy all foods from pickled eggs to haggis, chips and sushi. But although they used to go to the potty at the same time, they now have very different bowel habits and loo-visiting regimes, despite having identical diets and lifestyles. We tested their microbiomes and they were very unalike in many of the normal species. On average they shared only a minority of the same microbes, the same as unrelated people. This suggests that the antibiotics had removed the genetic similarity they started life with.

We now use antibiotics like sweets

In the US alone over 250 million antibiotic courses are prescribed every year, and recent studies in the UK show that despite grave warnings of overuse in general practice, rates are still increasing. Back in 1999 GPs were warned to cut back on prescriptions for mild infections and viruses. The warnings went unheeded – in fact, the situation got much worse. By 2011, rates of use had increased by 40 per cent, and the average GP was giving antibiotics to over half of all patients presenting with coughs or colds. These infections are due to viruses that antibiotics don't touch. One in ten GPs were even more cavalier, giving them to 97 per cent of their patients, presumably to keep them happy or to get them out of the surgery more quickly.

Global use of antibiotics has been increasing over the last thirty years in virtually every country that has data. Forty per cent of the prescriptions handed over are completely ineffectual, for the reasons just given.[1] All countries overuse antibiotics, but those with well-controlled centralised health care systems like Sweden and Denmark use the least – proportionally, half of what America uses. They also use more narrow-spectrum drugs, which select their targets more subtly and cause less microbial damage without any increase in problems.[2]

Even in rare cases where bacteria are definitely involved, independent overviews show that the benefits of antibiotics are minimal.

Treating sore throats or sinus infections early, for example, reduces symptoms by an average of only one day. This might seem worthwhile for some people, but that would be true only if there were no downside.

Treatments to die for

When Arun was two he needed antibiotics for the first time. His mother thought nothing of it as she had been on antibiotics countless times herself as a child, and believed them to be safe and effective. Serious side effects never crossed her mind.

Their ordeal started with what appeared to be a mosquito bite after Arun had been playing outside one evening. When he came in she put anti-itch insect-bite medication on the spot, and put him to bed. By the next day the bite was red and looked infected, and appeared to be spreading slightly up his leg. It was too late in the day to see his own doctor, so his mother took him to A&E at the local hospital where they gave him a shot of Ceftriaxone, which is a powerful type of cephalosporin, an antibiotic widely used to fight multiple unknown bacteria. To be extra sure, he was also given another antibiotic, Bactrim (a combination of two antibiotics), as a syrup, and his mother was told to continue with this course of treatment for ten days.

Shortly after he started taking the antibiotics his leg started healing, but he developed bad diarrhoea. His mother wasn't too worried as she knew this was a common side effect of antibiotics, but the diarrhoea was severe and persistent and then she noticed blood in his stools. She took him to the doctor's where a stool test was done, and was later informed that he had tested positive for Clostridium difficile (C. diff) and now had a condition called pseudo-membranous colitis, a nasty inflammatory disorder affecting the colon. His doctor then put him on yet another antibiotic, Flagyl, the first-line treatment for the colitis caused by C. diff. During the first few days of the new treatment his stomach problems improved greatly, but on the last day of the course the original symptoms returned. The doctor prescribed the same course again, and the same pattern recurred.

'At this point we were referred to a paediatric gastroenterologist,'

Arun's mother said, 'but had to wait an agonising week to get an appointment. I was terrified and called my doctor because I could not imagine waiting that long given his dreadful symptoms. He was losing weight rapidly and was visibly ill. In the course of researching his condition I learned about a rare but serious complication that can cause the colon to rupture, and is often fatal. Our own doctor said there was nothing more that he could do, and told me to take my son to the children's hospital if I became too worried. I was frantic and didn't sleep for two days. Then, miraculously, he got better – I'll never know why – but he could easily have died. People need to know of the harm that antibiotics can do.'

Others are not so lucky. Up to half of young children who contract this form of colitis can die, as their gut is so ravaged by the antibiotics that the immune system and gut barrier totally fail. This often begins when a round of antibiotics depletes the natural bacterial flora in the colon, reducing the diversity and power of the usual protective community and allowing a certain aggressive, or pathogenic, type of C. diff bacteria to flourish, replicate and eventually to totally dominate the gut in large numbers. Although horrific, these cases are rare, occurring once in every 10,000 uses of antibiotics. The risks are higher in bottle-fed infants as they lack the diverse and healthy bacteria like the bifidobacteria that come free with prebiotic-loaded breast milk. These extra microbes give them extra infection-fighting properties, as well as reducing allergies.[3]

Because we now prescribe millions of antibiotic courses annually, mostly of the non-specific broad-spectrum type that kill not only the likely pathogenic bacteria but everything else in their path, these severe C. diff cases, along with antibiotic resistance in general, are increasing. They serve to remind us of the dire and often hidden consequences of overuse, especially for seemingly trivial reasons.

Sterile births and future problems

Remember that the first three years of our lives are the most important in forming our core set of gut microbes whose job it is to maintain our health. Sadly, many drugs are given around birth without any thought for the poor microbes. Giving pregnant mothers antibiotics for mild urinary tract infections is commonplace, and for the last thirty years powerful wide-ranging intravenous antibiotics like cephalosporins have been routinely prescribed for mothers just before caesarean section operations so as to reduce the 1 to 3 per cent risk of post-op infections. This drug crosses the placenta to the baby and affects the breast milk, and may have even worse effects.[4]

I am all in favour of some C-sections. My life was saved by an emergency C-section when the blood supply to my mother's placenta suddenly packed up. I was very premature, a thirty-week four-pound weakling, and a few years earlier I would not have survived. I expressed my gratitude when, rather spookily, twenty-five years later I was delivering babies in a small hospital near Colchester and found myself holding a retractor for the very surgeon who had got out of bed to save me at 3 a.m. many years before. I had found his name on some old birth records – but strangely, he didn't recognise me.

Life-saving emergency operations are one thing, elective operations another.

In Europe there are major differences between countries. Italy, unsurprisingly, led the league table in 2010 with 38 per cent of births being caesareans, and other countries, such as Greece, may be over 50 per cent. Rates in all countries have increased since 2000, and there is a rough north–south European gradient, the UK in the middle with 23 per cent. The lowest C-section rates in Europe – and probably in the developed world, where the situation has hardly changed since the 1980s – are found in the 'deprived' countries, led by the Netherlands at 14 per cent, followed closely by the Nordic countries. These probably represent the sensible target levels.

In the US in 1968 only 1 in 25 births ended in an operation, but now it is nearly 1 in 3 and over 1.3 million operations are performed every year.[5] But the figures vary tenfold between areas, from as low as 7 per cent in some towns to 50 per cent in New York City and 60 per cent in Puerto Rico.[6] The operations are most frequently performed on women at the lowest risk of birth problems, as well as in poor populations that can least afford it such as Brazil (45 per cent in public hospitals) and Mexico (37 per cent). The epidemic has even reached China, where the majority of births are C-sections.[7] Cosmetic, financial and cultural reasons are likely to play a role in these differences, but more likely is the key role of the doctors, who no longer have to wake at 2 a.m. and can improve their golf handicaps.

Double trouble

Maria was thirty years old and already had one child and was about to deliver twins. She worked in a hospital, so knew the routine. After some discussion, it had been decided that she would have a C-section at thirty-seven weeks. The day had finally come. She had been fasting, and had been given a small enema to keep everything clean. The operating theatre was full of staff, plus her nervous husband standing there awkwardly in a sterile theatre outfit complete with mask. He and Maria were screened off from the obstetric team down the other end.

She was given a light anaesthetic and an epidural injection into her spine. Maria was relieved when told finally that she had two healthy boys, and glimpsed them briefly before they were whisked away. Thirty minutes later after being stitched up, she was handed her two tiny bundles for the first time. They were smaller than average, but not dangerously so – both were over 2 kg. They looked strikingly similar to each other.

She started breast-feeding and both boys slowly gained weight. At home a week later, the story was different. Juan was not gaining as much weight and seemed to be crying more than Marco. After two months the strain of breast-feeding persuaded Maria to introduce formula milk, which both boys accepted. But still Juan was lagging behind his twin and having more sleepless nights, with episodes of colic. By the age of two, Marco was a chubby, happy baby but Juan was a skinnier, unhappy version. Maria took Juan to the paediatrician several times and was told not to worry.

Eventually it was suggested she try Juan on soy milk as he might be intolerant to lactose. This initially helped him gain weight, but he developed a few strange minor allergies. The family arranged a DNA test for the twins, who were by now markedly different in size. The test confirmed they were identical. The reason for the discrepancy in weight, given that they were genetic clones and had been treated the same way, puzzled both the doctors and the family.

The gut microbiomes of the twins, as for all newborns, started out as blank slates. As the guts of twins become colonised by microbes from both the mother and the environment, they develop gut communities that resemble each other more than do those of fraternal twins or unrelated individuals, but they are not identical. Twins born by C-section, however, are on average more different in their microbes than those born naturally – and sometimes for strange reasons. For example, a few subtle differences in the way they are handled after birth could have a dramatic impact.

Back to Juan and Marco. The actual operation itself was pretty sterile. But when the twins were separated from the placenta they were each handed to a different nurse. The nurses may have looked very clean, but despite scrubbing up and wearing special clothing they were teeming with microbes, constantly shedding them from their hair, their skin and their mouths. In the course of weighing, tagging and cleaning, they were depositing new and different microbes on the fertile ground of the new babies. These 'foreign' microbes that found their way to the twins' mouths and guts were not what evolution had intended. So before they were reintroduced to their mother they already had a distinct microbial signature that not only determined what foods they would tolerate but would shape the rest of their lives.

In normal deliveries the infant gut is first seeded by microbes from the birth canal, including vaginal, urinary and gut microbes, followed by others from the skin. This produces a rich diversity of starting material for the crucial first three years, when, as noted earlier, the character and complex interactions in the gut are formed. These microbial communities are key to our normal development, and particularly in training our immune system, which has to learn from scratch. Vaginal microbes in particular change dramatically during pregnancy in preparation for the birthing process, and when they are altered can trigger early labour. Babies born via C-section are extracted before they can be exposed to the normal microbes from the traditional evolutionary route.

Studies have shown major differences, that happen in the first twenty-four hours, in the gut microbes of C-section babies compared with those in the guts of babies born vaginally. The most obvious difference is the lack of helpful vaginal microbes like lactobacillus, which are replaced by skin microbes like staphylococcus (staph, the cause of most minor skin infections) and corynebacteria.[8] We have seen that these don't all come from the mother, and that sectioned babies get most of their bacteria from the skin of strangers in the operating suite, or sometimes from the dad if he hasn't fainted and been removed.

So within the first few hours, changes in these keystone species have been made, and will stay different for at least three years, and maybe for life. The guts of C-section babies are also more resistant to subsequent colonisation by friendly bacteria like lactobacillus and bifidobacteria, even when they start normal breast feeding.[9]

C-Section babies have more allergies

Just as important, C-section babies, with their disrupted gut microbiome, have also a disrupted immune system that brings with it a greater risk of later immune problems such as coeliac disease and allergies, especially to food.[10] Nearly all the epidemiological studies (observational, not trials) published show an average 20 per cent increase in food allergies and asthma in these babies.[11] Most studies show the risk is greatest from mothers who have allergies themselves – in such cases the risk may be sevenfold. Most studies uncover similar risks in routine and emergency operations, which makes bias factors less likely and the results more believable.

So this amazing innovation – the opportunity to avoid natural labour and delivery – that one in three humans is now exposed to is messing with powerful evolutionary forces that we hadn't considered before.

Apart from banning C-sections, which would be unlikely to work, are there any realistic alternatives?

When Rob Knight of the American Gut Project carried out an earlier project comparing C-section births, he suggested something rather strange, on the face of it, to his pregnant wife. If she needed a C-section, he said, could he try to rebalance her natural state? She agreed, and as it turned out she needed a C-section for their daughter. Before she was anaesthetised Rob helped put a large swab between her legs and rubbed it around her bottom. When the healthy baby girl was extracted by the surgeons, Rob immediately wiped the swab across the face, mouth and eyes of his baby for a few seconds, trying to recreate what nature would have done.

After three years she is none the worse for this and has a natural-looking microbiome. Although her mother's family are very allergic she has no reported allergies so far, and only one infection that needed antibiotics, a bout of staph laryngitis. I have heard that this novel procedure already goes on unofficially in some Nordic hospitals. Rob and Maria Dominguez-Bello have started a trial of 'vaginal inoculations', as the process is now called, in Puerto Rico for women having elective sections, and will be following up the babies long term to see if they can be 'normalised' and have their allergies reduced. Other studies are exploring if you can compensate by adding a probiotic that is universal to all naturally born infants called *Bifidobacterium infantis*.

Nature has perfected the transfer of helpful nutrients and immune signals from the mother to the next generation not just via her genes but also from her microbes, which are finely tuned by what she eats during pregnancy. The dramatic increases in allergies in the last two generations can be explained by the diminishing diversity of microbes in babies, which is altering their immune systems in ways we don't yet fully understand. Once the baby is born, the odds are very high that he or she will be given antibiotics within the first three years; most countries now report averages of one to three courses during this time. When this happens, the delicate balance of the carefully forming microbial communities is upset and may never recover.

Of the eight most common adverse effects of prescribed medications in the USA and Europe, five are attributable to antibiotics. When the average American child reaches adulthood, he has already had seventeen courses of these drugs, and as in the UK most of them are unnecessary. Many young kids, of course, have even more than 'the average', which as well as reducing their immunity to other infections can have other nasty side effects.

Paediatricians working in developing countries have known for many years that chronic infections stunt child growth, which explains the link between poverty and short stature. A recent review of ten trials of long-term antibiotics given to young children found that they did indeed increase height, by half a centimetre a year, but they had an even greater effect on weight gain.[12] In these young Africans and South Americans antibiotics are helpful overall and can reduce malnutrition, presumably killing off many harmful microbes in the process, but this benefit is unlikely to be relevant to the West.

Antibiotics and obesity

Marty Blaser is a New York microbiologist who was one of the first to realise the potential long-term dangers of antibiotics and of our misguided attempts to eradicate microbes without thinking of the side effects. I first heard him talk in 2009 at a genetics meeting on Long Island, New York, and he convinced me of the reality of those dangers. He has now written an excellent book on the subject.[13]

He had seen, like many of us, a government study of the obesity changes over the last twenty-one years in the American states. The results were presented visually as coloured maps, changing over time, that you can watch like a horror movie.[14] The colours change from a light blue (less than 10 per cent obese) in 1985 to dark blue, brown, then red (over 25 per cent obese) – just like the representation of a plague. In 1989 no state had more than 14 per cent of its population obese. By 2010 no state had less than 20 per cent – even the healthiest state, Colorado. The southern US has the highest rates and the west the least. Over a third (34 per cent) of US adults are now obese.

Explaining these changes is not easy. But there are indicators. In 2010 rates of antibiotic use in the same states were published too. The results again showed large differences across the country, which couldn't be explained by illness or demographics. Amazingly, the same map colours for each state overlapped for antibiotic use and for obesity. The southern states, with the highest rates of antibiotic use, were also the states with the highest obesity rates. California and Oregon showed the lowest use of antibiotics (on average, 30 per cent less than other states), and it was these states that were relatively protected from obesity.

Now we are well aware that studies of national observations such as these can easily mislead. You could, for example, have a similar

US map correlating obesity with the use of Facebook or body piercing. So the findings of our two studies were far from proof. Some replication to confirm the antibiotic–obesity hypothesis was needed. The first opportunity came using data from the Avon Longitudinal Study of Parents and Children that I often work with. This follows up 12,000 kids from birth in Bristol, using carefully collected measurements and medical records.[15] In this study, exposure to antibiotics in the first six months of life significantly increased – by 22 per cent – the children's amount of fat and their overall risk of obesity over the next three years. In a later study the effect of antibiotics was found to be weaker, and there was no effect of other medications. This mirrored studies in a Danish birth cohort in which an effect was found between antibiotic use in the first six months and subsequent weight at the age of seven.[16]

A much larger US study has recently reported the results in 64,000 children. The researchers were able to compare the type of antibiotic used and the precise timing.[17] Nearly 70 per cent of kids from Pennsylvania had taken an average of two antibiotic courses before they reached the age of two. They found that broad-spectrum antibiotics given before that age increased the risk of obesity as a toddler by an average of 11 per cent, with greater risk if the drugs were given early.

By contrast, narrow-spectrum targeted antibiotics that kill a more limited range of microbes had no clear effect, and neither did common infections. These 'epidemiological' results, although supportive, are not conclusive and could have been caused by some other bias factors, such as children who take antibiotics being different or more susceptible in some other way. So Marty Blaser and his team took this one step further and tested his antibiotic theory in mice.

To mimic the effects of antibiotics on babies in the first three years, they used a group of lab mice pups divided into groups with and without antibiotics. He gave one set three shots of antibiotics for five days at the equivalent doses that babies are given for throat or ear infections. The antibiotic group grew much faster and had a disrupted and less diverse microbiome.[18] The results were clear and dramatic: in the antibiotic-treated pups there were significant increases in weight and body-fat levels, and the effects were greatest in the mice that had also been fed high-fat diets.

Unless especially lucky, most of us born in the last sixty years won't have escaped antibiotics as young children, or high-fat diets at some time in our lives, and could potentially be suffering the same effects as these mice. I asked our 10,000 adult twins from around the UK if there were any among them who had never taken antibiotics so that we could study them and their microbes. Sadly, we couldn't find a single individual. Even if you escaped antibiotics as a child, like me you may not have avoided being born by C-section. After adjusting for other factors, a meta-analysis showed that if you were born by C-section and didn't get the magic bottom swab treatment, your risk of obesity probably increases by 20 per cent, likely in my mind to be due to microbes.[19]

Animal junkies

Most antibiotics produced and sold are not for humans. In Europe around 70 per cent of antibiotics are destined for agriculture, with again big differences between neighbouring countries in their usage. In the US, around 80 per cent of all antibiotics used now are for farming. These are enormous quantities – some 13 million kilograms in 2011 compared to only 50 kg in the 1950s.[20] These poor animals must have a lot of sore throats, you might think. In fact the antibiotics are used for other reasons.

Following the war years and into the 1960s, scientists were playing around with ways of making animals grow faster.[21] What they eventually discovered, after lots of trial and error, was that adding continuous low doses of antibiotics to the feed of almost any animal increased dramatically its rate of growth, enabling it to be brought to market sooner and thus more cheaply – this was so-called feed efficiency. Moreover, the earlier in their lives you started the 'special' feed, the better the results. As antibiotics became cheaper, this made financial sense for the industry. And if this worked so consistently in cattle and poultry, why not in humans?

American farms no longer resemble farms as we know them. US farming today is famous for its enormous industrial-scale feeding stations that they call CAFOs (concentrated animal feeding operations), which can contain up to 500,000 chickens or pigs and up to 50,000 cattle. Cattle are bred super-fast and go from calf to slaughter in around 14 months, by which time they already weigh on average an enormous 545 kg.[22] The young cattle are rapidly weaned off natural hay and grass and trained to eat mass-produced corn laced with low-dose antibiotics. The corn is cheap, subsidised and plentiful, and grown in huge pesticide-laden cornfields amounting to a total area equivalent in size to the whole of the UK. Because of their new artificial diet that makes them ill, the overcrowding, the lack of fresh air and inbreeding, these animals are prone to infectious epidemics, so, paradoxically, they benefit from the antibiotics.

The use of only a few antibiotics has been banned from such widespread industrial farming. The US Department of Agriculture has been slow to seriously interfere in this lucrative market. In 1998, realising the potential for antibiotics getting into the human food chain and causing drug resistance, the more eco-sensitive European Union banned the feeding of certain antibiotics, valuable to human health, to animals. Then in 2006 they banned all drugs, including antibiotics, used for growth promotion purposes.

This should mean most meat is free of antibiotics in Europe. Sadly this is not the case: illegal use in feed is still rife, as scandals in the Netherlands have shown.[23] EU farmers are still legally allowed to use antibiotics when problems occur, and they do this regularly and often using extremely high dosages. Although the EU is trying to restrict which drugs can be used, in reality it has little control. For a farmer with one infected animal in a herd it is cheaper to treat all five hundred than isolate the animal and wait and see. Such huge amounts of antibiotics in both the food chain and the environment lead to increasing microbial resistance, requiring ever stronger antibiotics for the animals, and subsequently for us humans too.

Livestock producers outside Europe fail to abide even by these liberal rules. Furthermore, the European Union imports a lot of outside produce so you don't always know where your processed meat product comes from, or even if it's from the same animal declared on the packaging, as the European horsemeat-lasagne scandals revealed.

Over a third of our fish is intensively farmed, whether it's salmon from Norway or Chile or giant, fast-growing tiger prawns, grown in their billions in deforested mangrove swamps in Thailand or Vietnam. Antibiotics are now being used in fish farms in ever-increasing amounts, and most of these large suppliers are outside of European or American control. The worse the conditions the fish are kept in, the greater the number of tons of antibiotics that are needed. It has been estimated that over 75 per cent of antibiotics given to farmed fish pass through the cages to other wild fish swimming locally, like cod, and thus get into the food chain that way.[24]

Can we avoid antibiotics?

So if you are a meat or fish eater you will most likely be ingesting antibiotics with your steak, pork or salmon. Although it is illegal, in many countries small amounts are often detectable in milk. Even if you are a strict vegan and don't believe in antibiotics, you're not safe. Particularly in the US, but in other countries as well, contaminated antibiotic-fed livestock manure is used to fertilise plants and vegetables that may end up on your plate. And our water supply is contaminated by the millions of tons of antibiotics flushed down sinks and toilets and by animal waste, and now contains many bacterial colonies possessing antibiotic resistance.

Water companies keep it quiet, but they have no way of monitoring or filtering out either the antibiotics or the resistant bacteria. Large amounts of antibiotics are found in US and European water-treatment plants as well as in reservoirs in rural areas.[25] Similar studies have been undertaken in rivers, lakes and reservoirs all over the world, and their findings are very similar.[26] The higher the amounts and diversity of the drugs, the more resistant genes were found.[27] So wherever you live and whatever you eat, through your water supply you are getting regular doses. Even bottled mineral water may not be safe, as most varieties tested contain bacteria that have been exposed to antibiotics and are resistant to many of them.[28]

The commercial agriculture industry and the government food and agriculture agencies say that these doses that get into our food chain are totally harmless. But what if these august bodies, with 'no conflicts of interest' and concern only for your well-being, are wrong? Could these small doses be harming us? Marty Blaser, again, decided to test it empirically. And what his lab found was that mice fed on even tiny sub-therapeutic doses of antibiotics over their lifespan (mirroring its effect in farming) went on to produce twice the weight and body fat of normal mice and their lipid metabolism had changed.[29] The gut microbe contents had shifted significantly; there were many more Bacteroidetes and Prevotella species and fewer lactobacilli.

When the antibiotics were stopped, the microbial composition returned towards that of the untreated group, although it stayed less diverse. But subsequently, even on the same diets, these ex-antibiotic users remained fatter throughout life. The results were always more startling if you added antibiotics to high-fat diets rather than to normal healthy mouse food. Blaser's lab also found that the immune systems of the antibiotic group was badly impaired. The microbial changes interfered with the normal signalling pathways, and the genes controlling the immune system lining the gut wall and keeping it healthy had been suppressed.

To prove that the results were attributable to the change in the gut microbes and not to some direct toxic effect of the drugs themselves, the research team transplanted the microbes from the guts of the antibiotic-treated mice to the germ-free sterile mice. This produced the same noticeable weight gain, showing conclusively that it was the reduction in the gut flora that was the problem, not the antibiotics. Whether the animals had been fed high- or low-dose antibiotics, both groups also showed increases in the natural gut hormones associated with obesity, such as leptin and a hunger-inducing gut hormone known as PYY, released after signals from the brain that reduce food transit time and permit greater extraction of calories from all kinds of food. This reminds us of the important gut–brain interactions that occur all the time.

Modern babies are facing an onslaught of antibiotics, whether via injections given to the mother before a caesarean section, via short courses for minor infections, or via the mother's breast milk. And add to this the low-level contamination of tap water and food that we still don't know the effects of. Antibiotics could be the root of many

unrelated and unexpected health problems, such as the recent finding that treatment increases the risk of malaria spread and infection
by favouring the uptake of the microbe plasmodia by mosquitoes.[30]
Antibiotics may be the missing factor, or certainly one of them,
explaining the current childhood-rooted obesity epidemic. The dim-
inution in our microbes and our processed, sugary, fatty diets have
joined forces to produce the perfect obesity storm.

What's more, as we become fat and pass our highly selected
fat-loving microbes on to our children, a vicious cycle starts: the next
generation is exposed to yet more antibiotics and has an even more
impoverished microbiome than we had. In other words, the problem
of the depleted microbiome is escalating in each generation. This
explains why the effects and trends we are observing are magnified in
children of obese mums, who themselves started life with a defective
microbiome.

Given that antibiotics are so difficult to escape from, is there a
solution to this mess? Transforming yourself into a New Age anti-
medicine organic vegan might give you and your family and your
microbes a slight edge, but a better bet would be community action
with a view to reducing the use of these drugs.

The benefits to our children would be greatest if doctors were not
pressured into prescribing them. Clearly, for emergencies you need
to seek help, but for a minor ailment try waiting an extra day or two
to see if it resolves itself. If we all started accepting that we sometimes
get ill and put up with a possible extra half-day of symptoms without
a prescription we would keep our microbes happy. Governments can
make a difference by targeting the worst doctors. This is how, between 2002 and 2006, France managed to stem the tide and reduce
its antibiotic prescribing for kids by 36 per cent.

If we really need the drugs we should be using modern genetics to
develop more targeted ones, not wiping out the whole microbiome
garden with the current medicines. As well as all eating less meat,
organic if we can afford it, we should be lobbying government to
reduce subsidies for industrial-scale antibiotic-dependent meat production. With antibiotic resistance increasing exponentially globally,
we will soon have no drugs left to treat serious infections and we
should be working seriously on alternatives. It could mean using

natural bacteria-killing viruses that are harmless to us. For this we would need to increase research funding so as to enable us to rapidly identify and eliminate the guilty species.

Are probiotics the cure?

Can yoghurt drinks with added acidophilus or bifidos, or other freeze-dried probiotic microbes, make any difference? As we already discussed, if you are very young, or old and severely ill, there is increasing evidence that they are beneficial.[31] For the rest of us, the reality is that although they do no harm there are as yet no good trials showing major benefits in humans. This is probably because we all have very different gut microbes to start with. So not knowing which microbes to replace, it's a lottery whether such yoghurt drinks will work for you.

Hopefully, the future will produce tailored probiotics for each person that will depend on all of us having our gut microbes measured routinely, which can of course be done.[32] Meanwhile, eating microbe-supporting prebiotic-rich foods (like artichokes, chicory, leeks and celeriac as well as fermented foods and yoghurts) while you are on antibiotics seems sensible, although we are still awaiting data to back this up. And since antibiotics and C-sections have been linked with a rise in allergies, some of the possible culprits being microbe-friendly foods, should we be restricting our diets further to prevent them?

18

Warning: May Contain Nuts

Fae had turned blue, her lips had blistered, her face was horribly swollen and she had stopped breathing. Her mother was screaming. And this was all happening at 30,000 feet up in the air. Fae was a happy four-year-old from Essex, who five minutes before had been playing with her sister. The family were returning to the UK on board a Ryanair flight from a sunny holiday in Tenerife. The little girl had a severe allergy problem, and the crew had reminded the passengers three times not to open any bags of nuts. One man, sitting four rows behind Fae and her family, took no notice of the warnings because he thought, like many of the others, that they were exaggerating the danger. Despite his neighbour trying to stop him, his apparently desperate need to tuck into his bag of mixed nuts to help him cope with the three-hour flight got the better of him.

A few minutes later, because of the powerful air-conditioning system that circulates food and dust particles in aeroplanes, some nut dust must have got into the air. Fae started scratching her cheeks, became red, then lost consciousness. Her mother rushed her to the front of the plane, away from the source of the nuts, but the damage was done. Fae's dad found the adrenalin pen they always carried, but in his stressed state and with his hands shaking he couldn't get it to work. Their daughter was dying in front of them.

The cabin crew hadn't been well-trained either, and were helpless until one of the passengers, an ambulance worker, dashed forward and gave the injection. Fae slowly came back to life, to the relief of the entire plane including the nut-loving man, who felt very guilty and was nearly beaten up by his fellow passengers. He was subsequently banned from flying with Ryanair for two years.[1]

Nuts were part of the diet of many of our ancestors, are a common ingredient in many contemporary cuisines and a key part of the healthy Mediterranean diet we discussed earlier. There are many types of edible nuts, which contain a mixture of mainly unsaturated fats, some protein and polyphenols. For the average European and American they provide about a fifth of the total antioxidant polyphenol content of the diet. Walnuts contain the most, with over 20 polyphenol chemicals.[2] One 30 gram serving of walnuts is equivalent to the average amounts of polyphenols from the same amount of fruit and vegetables combined. For peanut-butter-loving Americans, peanuts supply two-thirds of their antioxidant intake, making up for the sugar and calories that usually go with it. Generally, when you roast nuts you get 15 per cent more antioxidant polyphenols, but this varies widely between different nuts. There is a current vogue for soaking your nuts (not that kind) before eating, so as to release nutrients and eliminate toxins. This is not harmful – it makes sense with some beans – but by now you should be wary of anyone telling you to eliminate imaginary toxins in real food. Your gut microbes are responsible for making the nutrients available even from the toughest nuts.

In the 1980s, because of their high cholesterol and fat content nuts were especially targeted as bad for us, and I too used to think they weren't healthy. But increasingly, studies show that if not over-salted they can suppress appetite, and they've been shown in prospective studies to help reduce weight and improve blood lipid levels.[3] An extra 30 grams of mixed nuts, eaten raw (and unsoaked) in the famous PREDIMED randomised diet study, produced significant advantages over the low-fat heart-disease diet and was nearly as good as the extra olive oil group.[4]

Peanuts are technically legumes, not nuts, but the same rules apply. Americans spend nearly $800 million a year on peanut butter which, despite its bad reputation, contains vitamins, protein, fat and fibre and, if not over-processed, may even have some modest benefits for the heart. Nuts alone are still better at reducing mortality according to a large Dutch study.[5]

There are many other chemicals in nuts we know little about, such as cachexins which may help with losing weight. If you don't cover these complex bundles of food in sugar or salt, nuts overall are good for us. So when did they become associated with DANGER, as blazoned on food labels and in restaurant menus, and become deadly weapons on aeroplanes? Have nuts changed, or have we?

Food allergies – a modern phenomenon?

The first ever medical mention of a food allergy (eggs and cow's milk) occurred the same year that the *Titanic* sank, 1912.[6] Your chances of getting an allergy were, however, much less than hitting an iceberg. A hundred years ago, if a patient presented to a physician with a food allergy his eyes would light up. He could write up the patient's story in articles and regale his colleagues with its exotic rarity, write a popular book and become a travelling celebrity. By contrast, the first food allergy case to appear in a modern medical journal happened the same year that man landed on the moon, in 1969.[7]

Stories of children like Fae collapsing dramatically in mid-air are becoming increasingly common, as are the numbers of people who are severely allergic with potentially fatal anaphylactic reactions, although the number of annual child fatalities is strangely but luckily tiny (single digits for the UK). Nut-free schools started in California and are being copied elsewhere, where worried parents can send their children. If the trends were to continue and more pressured airlines like British Airways threaten to ban nuts, a bag of peanuts could be seen as a weapon of mass destruction, and special flights would be needed to transport allergic kids on holiday as it would be impossible to totally protect them against some tiny allergen in the air. But is this yet another urban myth?

I asked several leading allergy consultants in top London hospitals about this amazing Ryanair case, which had been a major topic of conversation and prompted calls for the banning of in-flight peanuts. The results were unexpected. All said they would have happily eaten a bag of peanuts next to the young allergic girl. According to them a piece of nut would have to have flicked directly into her mouth for any reaction to occur. Most allergenic foods don't transmit in the air or in dust, and this includes the main peanut allergen protein ARAh2. The only explanation the allergy specialists have is that Fae's problems were set off by some other unrelated triggering allergen that found its way directly to her mouth.

These doctors, who perform hundreds of allergy tests monthly on high-risk kids, say this kind of severe reaction is very rare and never occurs with peanuts transmitted by air. One exception to this is fish allergy because it's the fish odour that contains the allergenic protein. This illustrates the power of rare-disease pressure groups and of the media in transmitting myths that can be long-lasting and have important sociological effects. The next time on a plane you get asked not to eat nuts because of some child with an allergy and understandably anxious parents, how will you react?

Notwithstanding the considerable hype and scaremongering, food allergy is definitely on the increase. Food intolerance is also increasing, but this is very different and much harder to define. Allergy is an obvious and immediate reaction to food, producing swelling, redness, numbness and often loss of breathing or consciousness. You can't have a slight allergy any more than you can have a slight case of cancer or death. Intolerance, however, is usually described as involving bloating, nausea, gut pain, diarrhoea or constipation after eating. We don't know whether all these have really increased recently, or if it's that they existed before but people didn't previously report them as an illness. Many of my patients with rheumatoid arthritis believed that different foods had caused their disease or worsened their symptoms. I always told them that rather than spend their money on pseudo-scientific allergy tests on their hair samples and bio-rhythms, they should try an exclusion diet of mineral water and vegetables for two weeks to see if they improved. No one did. The real rate of food allergies being related to diseases like this is less than 1 per cent.

But allergies are a real enough modern disease and the list of possible triggers is endless. These can range from one of the commonest, allergy to nickel – which in our set of middle-aged female twins affects one in five – to rarer ones like allergy to shrimps (one in fifty), bananas or tomatoes, and to even rarer ones as when people are allergic to the coating of pills or to sunlight (one in a thousand). Rarer still are people allergic to water (the condition has a name, 'aquagenic urticaria'); these individuals can be allergic to their own tears or their husbands' kisses. It's a very distressing condition, but a good way to get out of doing the dishes. Some strange combination allergies are also appearing nowadays: one girl has an allergic shock only if she eats apples that are contaminated with birch pollen.

Protecting your baby

Australia has one of the most allergic populations in the world, for reasons we don't understand, but it's unlikely to be due to pollution. One in fifty children there now suffers from peanut allergy (compared with one in eighty in the UK), and rates double every twenty years. The increase is most rapid in those under the age of five. The first five generations of Australians, who mostly came from Ireland and the British Isles, encountered many strange allergens when they arrived; but, seemingly, most had few problems until the last thirty years. Since my mother was raised there – and my brother and I went to school there for a few years in the 1960s – Australian life has changed dramatically.

This original Aussie outdoor, sporty, BBQ lifestyle, with its picnics, snakes, spiders, dirty feet and outside toilets (the 'dunny'), is no more. Children now rarely play outdoors, most stay inside with PlayStations and computers in clean, air-conditioned, vacuumed homes and eat clean and increasingly processed foods. Australia also has one of the highest child obesity rates in the world and, contrary to its image, few kids play sports. Drinking beer and watching sports have, however, survived pretty well.

When do kids become allergic? The latest research shows that babies are sensitised to the specific proteins in food (allergens) not only when they encounter the food as an infant but *really* early in life – in the womb.[8] The traditional medical advice for worried mothers was to avoid certain foods such as French cheese or salamis, and there is a current tendency to progressively restrict what pregnant mothers eat because of risks of rare infections or allergies. Often the advice given is defensive and not evidence-based – and just at the time when the mother needs a healthy and diverse diet and when her body usually tells her what it needs, anyway. Specifically, women are told to avoid eating peanuts in pregnancy and hope for the best. Recent studies have shown, however, that the opposite is true: that women snacking on peanuts during pregnancy are actually much less likely to give birth to nut-allergic kids than those who abstain.[9]

Having a diverse microbiome when you are a newborn infant appears to be essential in reducing the future risk of allergies.[10] The source of this diversity comes from breast milk assisted by a healthy maternal diet, late weaning or a bit less household cleanliness. By contrast, it's a microbiome *reduced* in richness and diversity that is usually associated with allergies.[11] Allergies are more common in bottle-fed babies with weakened immune systems; the current theory, as already mentioned, is that a healthy and diverse microbiome keeps the gut immune system stimulated and in a constant state of readiness so that there will be no overreaction to strange proteins.[12]

Hygeine hysteria

The babies of mothers who run ultra-clean homes and have super-clean babies are most at risk of allergies. A study recently found that babies who had their rubber dummies sucked clean by a parent then popped back into their mouths had considerably fewer allergies than those with parents dutifully replacing hygienic sterile dummies.[13] The old-fashioned practice of mothers pre-chewing their baby's food, which is rare in the West nowadays, served both to break down tough starchy foods and meats and to transmit a wide range of helpful microbes via saliva. Licking babies is common in most mammals and in some human cultures, and of course kissing is pretty universal.

The Hygiene Hypothesis is an idea you may have heard of. It was developed by a colleague I trained with in epidemiology, David Strachan, whose interest was sparked when he was looking at the national data of children followed up from birth for asthma and eczema. He found a correlation between damp housing conditions and allergy in the UK.[14] But the link was not what we might intuitively have expected: the damp, poor conditions and overcrowded families were actually protective, even after adjusting for other possible sources of bias. The results were backed up and replicated in many other populations. Thus was the hypothesis that excess hygiene could lead to modern allergic diseases born.

People brought up in less sanitary conditions and regularly exposed to animals and worm infestations never seemed to develop asthma or food allergies. This was believed at first to be purely to do with the immune system, which needed to be stimulated by infections early in life to fine-tune its defences. This was how we had evolved over millions of years. Suddenly, from the 1960s onwards children were being brought up in increasingly hygienic environments without exposure to worms and dirt and a variety of mild diseases that in earlier times had served to educate the immune system. Thus the richer a country became and the more they protected themselves against the natural world, the more allergies would ensue.

We have probably seen the worst of the asthma epidemic, which was at its height in the 1980s and 1990s, but as these levels are falling we are encountering a massive increase in severe food and skin allergies. Unlike asthma, they don't seem to disappear in adulthood. One in twenty children is allergic to peanuts, milk and other foods; the rate has increased by around 3 per cent annually over the last twenty years. Gluten allergy, too, is seen more often than before. Skin-prick and patch testing offer one way of quantifying how common allergies really are, without depending on unreliable questions. In recent US surveys 54 per cent of children show evidence of a mild allergy to something. We tested our older UK adult twins from both rural and urban areas, and one in three had a positive skin-patch test suggesting a potential allergy.

However, some groups in the US are relatively well protected from the allergy epidemic. Researchers from Indiana who studied the local Amish found only 7 per cent of Amish children had positive results from skin-prick tests, which is six times fewer than genetically similar Swiss children.[15] The Amish way of living hasn't changed much since they left Berne in Switzerland in the seventeenth century. All kids are raised communally and taught to walk and milk cows in dusty barns full of hay, straw, animal hair and manure. The workers that do most of the farming possess the most diverse sets of gut microbes, with large amounts of some species like Prevotella which, as we have seen, are rare in the rest of America but common in Africa.[16]

The Hygiene Hypothesis has stood the test of time so far, but now has to be adapted to our new knowledge of the importance of microbes. We need to remember that our gut microbes play a key role in training our immune systems. They do this via communication with the Treg cells in the gut walls, which are the main communicators and thermostats between what we eat and how our immune systems react.[17] High Treg levels are generally healthy as they suppress the immune system. It is no surprise, therefore, that kids born to mothers with food allergy already have low levels of Treg cells at birth, because of a combination of their parents' genes and their restrictive diets.[18]

One reason the Amish have such low rates of allergy is that they drink plenty of raw unpasteurised microbial milk. Similar studies have been carried out on European families. If we could design an Amish-style diet or probiotics specifically for mothers and their babies that would maximise the Tregs' chemical signals, we could start to reverse the food allergy epidemic.[19 20] Most of us were brought up with the idea that hygiene is top priority for healthy babies. But should we still be trying to keep our babies clean? The protection the Amish get from their natural but dirty environment comes not just from the dust and hair from all the animals around them, but from the trillions of microbes that live off them. European kids brought up near farms have less asthma and fewer allergies.

Not all farms nowadays are the same. In the US, living near giant CAFO feeding stations can reduce food allergies but also increase asthma.[21] One study that allowed babies to play outside and roll about in the mud and dirt for most of the day showed fewer allergies and immune problems in their subjects, and their guts contained more friendly lactobacilli than babies regularly washed and kept inside. I'm talking about baby piglets here, by the way. But in terms of microbes, genes and our health we are very close to pigs.[22]

Many anxious mothers of children with allergies are racked with guilt, trying to protect them from attacks from deadly allergens in dust and animal hair. They try very hard to keep their houses in a state more appropriate to a sterile laboratory. Others are so worried about tiny particles of nuts, gluten, wheat or eggs getting into their food that eating becomes as relaxing as taking part in a feast at the Borgias'. These fears are understandable, given that nut allergies can cause death on rare occasions. But studies show that keeping your house more like a farm and having pets and even pigs in the house can reduce, not increase, allergy rates. Some of the health advantages of keeping pets in terms of human longevity, allergy and even avoiding depression come from not only sharing their hairs and dirt, but more intimately from their diverse range of gut microbes.[23] Embracing dirt and diversity as our friends, and abandoning the notion that our food intolerances are in fact major medical problems, is a tall order, but could be crucial for the next generation.

19

Best-before Date

An amazing amount of edible food in our modern world is thrown away for no good reason. Estimates are that in the West between 30 and 50 per cent of most families' food shopping is discarded. In some cases this is understandable. Some of the fruit and vegetables that I always buy too much of in my bid to be healthy goes soft and mouldy, attracts insects and ends up in the bin. But this is the exception.

Many people now treat products past their best-before dates as deadly, or at least likely to cause food poisoning. The common myth is that microbes eventually take over the product and that ingesting the resulting mix of the two is likely to be toxic. Few people realise that these dates are just estimates of food quality, not of safety.

There are clearly real things to steer clear of, particularly when it comes to raw meat like mass-produced chicken, where salmonella or campylobacter can take over and potentially cause stomach infections; but if the item has been refrigerated this is not linked to length of storage. Most infections occur well before you buy the product. When supermarket foods age they may lose taste and structure and their microbe composition may change, but they are not harmful.

Good cheese, as we discussed, is already overrun with bacteria and fungi, but scraping off the mouldy crust is perfectly fine. The same goes for jams, yoghurt and pickles. Even preservatives like vinegar or olive oils which don't go off currently have best-before and use-by dates. I haven't done a scientific survey, but I have yet to meet anyone who got ill from anything in their own fridge. Strangely, most infections come from eating out, where you might have thought standards of hygiene were stricter. Luckily, food infections, which are usually due to meats and eggs, are steadily decreasing in most countries and are a quarter of the levels of twenty-five years ago.[1]

The sell-by dates on food labels were originally used to help supermarkets restock the shelves from the warehouse and improve efficiency, but they soon found that consumers liked reading them so that they could select the freshest products and reject those past the date. This led food companies, encouraged by governments, to add yet more confusing dates – best-before, sell-by, use-by etc. – as they quickly realised that this way more food got discarded, which increased sales. The whole business has now got out of hand, and many people are enraged at the billions of kilos of food thrown away by supermarkets as well as by consumers, who wrongly half-empty their fridges or cupboards every week purely on account of these labels.

In a recent survey of supermarket bosses, nearly all said that they regularly ate out-of-date products and regarded them as perfectly safe. While still keeping 'use-by' dates the EU has finally decided to phase out useless 'best-before' labels on totally safe products like rice and pasta in a bid to reduce confusion and waste. In the US the situation is worse, because each state still has different and confusing label regulations. The manufacturers are happy with the confusion, as the shorter the use-by dates the more people will throw away and restock. While the food authorities offer only a zero-risk policy, at home you can make your own choices. When faced with a mouldy loaf of bread, do you trim it and toast it, or bin it? Many different species of fungi love bread, so you could get very unlucky with a rare strain that might cause an allergic reaction. Mouldy nuts and soft fruits are also best thrown away because fungi can produce toxins, although it is probably fine to scrape and trim your cheese, salamis, hard fruits and vegetables if they look and smell OK.

We are also told never to reheat rice once cooked. This stems from old epidemics of food poisoning from the spores of a microbe (*Bacillus cereus*) that can hide in rice and are virtually unkillable. After heating, the spores can slowly come to life and produce a nasty toxin that causes a 24-hour tummy bug. Epidemics were frequent in the 1970s after fried rice was reheated by caterers but they are much rarer nowadays. Only two outbreaks, affecting a total of 220 people, were reported over the last five years in the UK. Estimates in the US suggest it affects 27,000 people a year and counts for 0.2 per cent of food-related illness. Like many people I know, I have been reheating rice from the fridge all of my life without problems, which suggests that if you are sensible the risk in the home is minimal.

Incidentally, we also throw away many expensive drugs unnecessarily; most prescribed medications are fine way beyond their use-by dates. Often the texture may change, but the active ingredients are still present. One study stringently tested 150 drugs and found 80 per cent were potent many years past their expiry date.[2] There are a few rare exceptions such as the antibiotic tetracycline, which loses its efficacy quickly, and liquids that can separate out. Most doctors' cupboards, like mine, are full of 'expired' medicines that they never throw away, and although they may lose up to 10 per cent of their potency there are no reports that I've been able to find of anyone suffering the effects of an expired drug. Charities like Médecins sans Frontières are collecting the returned medicines from pharmacies in a few Western countries, to use in the Third World, and the US have a program that will cautiously extend shelf lives.[3]

We should, and can, do a lot better than this. As far as food storage is concerned, our society needs to re-evaluate our ground rules and reassess the balance of risk. The minuscule health risks of consuming food or medicines that have lost their colour or texture need to be balanced against the implications of accepting zero risks: for a start, by throwing it in the bin we are certainly contributing to climate change.

Pesticides are an increasing potential problem for our microbes and they take many forms. The most popular is called glyphosate (or Roundup), which stops vegetables and fruit sprouting or going mouldy once developed. It was invented by Monsanto in the 1970s and is probably the most commonly used chemical for farming in the world. In 2013 over 1.7 million hectares of land in the UK was sprayed with it, and the majority of non-organic breads (especially wholemeal) tested contain glyphosate residues. Traces of it are found in the blood and urine of cattle and even in humans living in cities. Even at sub-toxic doses it could be adversely affecting human health and, like most chemicals, contains potential carcinogens.[4] We know it affects soil microbes, and much less is known about its effects on our gut microbes – but early studies suggest it is not good.[5] We may prefer to let our fruit and vegetables deteriorate and change colour

after a few days, rather than keep them chemically in suspended an-
imation with adverse effects on our microbes. While there is little
solid research on whether eating organic foods is better for us and
our microbes, there are studies showing levels of pesticides in our
bodies can be dramatically reduced within a week by switching to
organic produce.[6]

Should you peel your fruit and veg to avoid pesticides? This
is a modern dilemma. If you do, you lose much of the fibre and
nutrients, and it is likely futile as most of the pesticide will have
penetrated beneath the skin, so you may be better off giving them
a quick rinse. One subject we tested for the British Gut project had
very high levels of strange microbes in her gut that only normally
live in soil. It turned out she was a big organic food lover. So far they
seem to be passing through without causing her any problems and
are hopefully adding to her diversity.

Until we change our exaggerated fear of microbes we will find it
hard to make the right moral and health decisions. But while we
desperately need more reserach into organic products, we shouldn't
be complacent about increasing the levels of 'safe' chemicals in our
food just to make it look better for longer.

The Checkout

We have reached the end of the food label, and before we check out it is worth summarising the major diet myths, their fallacies, and what we can do to improve our own diets and our health.

The most dangerous of these myths is the notion that we all respond to food in the same way, that when we eat food or follow certain diets our bodies behave like the bodies of identical lab rats. They don't. We are all different. This is why the obsession with the limited view of nutrition and weight as calories-in versus calories-out is unhelpful and distracting. The truth is that each of us responds to food differently even if the food and the environment are identical. Remember the remarkable Canadian study I described at the beginning of the book that involved overfeeding lean student twins: although they were on identical food and exercise regimes, there was found to be a threefold difference between the weights individuals gained over the two months (one gained 4 kg and the other 13).

Our bodies vary entirely in how they respond to everything, from food to exercise to environment, and this variation affects how much fat we deposit and how much weight we gain as well as our food preferences. As we have discovered, the variation is due in part to our genes, but also to the different microbes that populate our guts. Certain groups and species are associated with protecting individuals against many diseases and against weight gain, while others increase susceptibility to these factors.

But despite the caveat that everyone is unique and will be affected differently, there are certain facts about diet that are unarguable: diets that are high in sugar and processed foods are bad for our microbes, and by extension for our health, and diets that are high in vegetables and fruits are good for both. The US writer on food and health Michael Pollan's simple seven-word message sums it up: 'Eat food, mostly plants, not too much.' And to modify another bit of his advice: 'Don't eat anything your great-grandmother's microbes wouldn't recognise as food.'

Our microbial diversity is declining every decade, which is definitely a bad thing and likely to be a major contributor to the modern epidemics of allergy, auto-immune disease, obesity and diabetes. It is clear that the more diverse your diet, the more diverse your microbes and the better your health at any age. But how easy is it to change old habits?

A new approach to food: never dine alone

In the five years that I spent researching and writing this book I have learnt a lot about myself and my relationship with food. I know that I would have to be at great risk of some fatal disease to give up cheese for ever and I would now miss yoghurt. My body seems to need some minimal amounts of the nutrients in meat protein – once or twice a month, to stave off vitamin deficiencies. I have worked out that a Mediterranean-style diet suits me very well. This may be because of my southern European genes or due to the chances I have to eat Mediterranean food in nice sunny places.

I am happy with yoghurt and fresh fruit in the mornings, plenty of varied salad, olive oil, and a wide range of other foods including fish in the evening. I have found that reducing amounts of over-refined carbs (white pasta, rice, potatoes etc.) is beneficial to my health and that replacing them with wholegrain varieties or legumes is fairly easy, although doing so has lengthened the time it takes me to prepare my meals. Since adding the extra fat from the olive oil, I found I didn't miss them so much anyway. A very pleasant surprise was that items I thought were naughty but that I enjoyed immensely, like strong coffee, dark chocolate, nuts, high fat yoghurt, wine and cheese, are actually likely to be healthy for me and my microbes. So to quote Barbara Kingsolver, who wrote the book *Animal, Vegetable, Miracle*, 'Food is the rare moral arena in which the ethical choice is generally the one more likely to make you groan with pleasure' – we might add that this is the choice more likely to make your microbes groan with pleasure too.

Diversify your diet and your microbes

Most diet plans contradict each other in their advice on meat versus vegetables and on low carbohydrates versus low fat. But the one thing that every diet guru, book and plan can agree on is that processed and fast foods are to be avoided. I've found that I can survive without eating processed foods apart from the occasional indulgence of soft liquorice or potato crisps. If you want to change your diet and to eat more healthily, say by consuming more fruit and particularly more vegetables, I've found that it's easier to give up something such as meat temporarily. This enables you to fill the space with other items that you may not have tried before and it's also a good excuse for saying no without upsetting hosts. Juicing was a new experience for me, and despite the washing-up afterwards, is quite a fun diversion at weekends. Certainly, the product tastes much better than it looks. It's a great way to use up spare vegetables in the fridge and increase the diversity in your diet. When on an alcohol fast for a few weeks in 'dry January' I found juicing was a great substitute.

As I learned to listen to my stomach more, I found that I was breaking with tradition and mealtime culture, which was very illuminating. I tried different ways and times of eating, which was also instructive. I learnt that on busy days I could skip breakfast or lunch and that intermittent fasting was much easier than I thought. Total fasting, like for my colonoscopy, was fine for short, well-defined periods. However, I realised this is much more difficult if you're not busy and distracted; for me it would have been tough to do at home at weekends. Just performing a fast of some kind is a useful exercise in itself; it shows that you can survive psychologically and that you're not risking death or coma by skipping a meal or a drink. It also allows your microbes to give your gut lining a useful spring clean.

I never tried a very low or zero carbohydrate and high protein option because my tolerance for meat is now limited, having been weaned off it. I am also now against deliberately avoiding any 'proper foods' altogether (like grains or pulses), especially when these are sources of nutrients or fibre. We need to eat more diverse foods, not fewer. Trying new foods is part of my new philosophy, as is eating as diverse a diet as I can within the constraints of a busy modern life. As a measure of good diversity I suggest aiming for ten to twenty food types, per week, that are good for your microbes. Rather than eating dull meals from the hospital canteen on a busy day, I usually now eat a piece of fruit and a handful of mixed nuts. I tell myself I will make up for this at dinner – which I do.

Exploring your own gut residents

Over the last year I have tested my own gut microbes dozens of times on different diets, although not as much as some colleagues, who collect samples every day of their lives (you wouldn't want to share a fridge with them). As I mentioned earlier, I have also set up the British Gut Project (www.britishgut.org), which is a copy of the larger American version that allows anyone to check out their own microbes by post and internet.[1] This service is provided for a small donation as long as the participants – we call them 'crowd-funders' – agree to share their data with the citizen scientists in the rest of the world. To participate, you simply wipe a cotton bud on some used toilet paper and send the sample to us in the post for sequencing. It is also strangely exciting when you receive your results and compare them to those of the rest of the world. Even I got emotional looking at my gut's contents. I found my profile was quite unusual at the big picture level: I had much lower levels of Bacteroidetes (18 per cent) than average and lots of Firmicutes, which was potentially a worry.

A few years ago it was believed this kind of ratio was a 'bad' profile that led to obesity and disease, but this was too simplistic a view. We have since found that many hundreds of species within the broad groups can have opposing effects, especially on obesity. These species, although we group them together, are very different from each other genetically – it's rather like comparing ourselves to starfish. Plotting all my gut microbes against the averages from the US, Venezuela and Malawi, I come out as a cross between North American and South American, which means I have a greater diversity than the average American (and my burger-eating son) – I turned out to be strangely similar to the microbial make-up of the writer Michael Pollan – but am not nearly as diverse as an average African.

Diversity or gene richness is a better indicator of health than the presence of any one microbial species. After my colonoscopy my prebiotic feast of vegetables managed to shift my microbes in a healthy direction, but not so much as to make anyone mistake me for a hunter-gatherer.

But what should you do if you want to change things more radically, even permanently? For a start, you could follow the lead of my unorthodox colleague Jeff Leach who decided he didn't like his Western gut microbes and wanted a change.

Time for a radical makeover

'I struggled to keep my legs in the air with my toes pointing towards what I thought was the faint outline of the Southern Cross rising in the evening sky. With my hands under my hips and butt perched against a large rock for support, I pedalled an imaginary upside down bicycle in the air to pass the time as I struggled to make sure my new gut ecosystem stayed put inside me. As the sun set over Lake Eyasi in Tanzania, nearly thirty minutes had passed since I had inserted a turkey baster into my bum and injected the feces of a Hadza man – a member of one of the last remaining hunter-gatherer tribes in the world – into the nether regions of my distal colon.'

This might seem a bit drastic, but as we saw earlier, Jeff is no wimp. He had tried to alter his microbial community by living and eating with the foraging Hadza for several days during their wet season. He consumed their baboon-dropping-infested water, their wild honey, tough tubers and the occasional zebra. His microbes had changed, but not as much as he would have liked. This is what led to his unorthodox experiment – standing on his head with the oversize turkey baster inserted into his rear end containing the 'donation' from a thirty-year-old male (tested for HIV and hepatitis). Jeff is a pioneer but others call him mad, particularly as he is not ill and is doing this 'for science'.

But Jeff's impulse to repopulate his gut using a donor is not as unique or crazy as it may seem. Around the world thousands of people who are seriously ill with C. diff infections have stool transplants (I still don't like the term 'faecal transplant'), mostly in specialised clinics with good success rates, although sometimes they need several attempts with beneficial species to eradicate the C. diff populations. Unlike Jeff's, most transplants are not conducted with turkey basters (which I found, incidentally, are also used for DIY artificial insemination). Bona fide transplant clinics use tubes passing into the colon as for colonoscopy, or for higher success rates they pass a thinner tube through the nose and down past the stomach.

For obvious reasons, researchers have perfected a more acceptable way. They freeze-dry a donor's stool after cryo-preserving it and insert it into capsules that are acid-resistant but dissolve in the colon releasing the microbes from their cold-induced slumber. These capsules (nicknamed 'crapsules') have similar rates of success to liquid methods for treating C. diff infection, and volunteers unsurprisingly prefer to swallow fifteen small 'organic' pills over two days rather than having plastic tubes inserted in different orifices with the tiny but nasty little thought of potential leakiness.[2]

Moreover, companies now offer fully tested super-healthy donor stools for transplants to reverse severe C. diff infections and treat some cases of colitis and severe bowel problems.[3] Like sperm and egg donors in the US, if you are healthy you get paid on delivery ($40). One not-for-profit company I visited, Openbiome in Boston, supplies nearly 500 US centres performing transplants from less than twenty donors. Other countries, such as the UK, are currently lagging behind, with fewer than ten transplant centres. All transplant methods have around 90 per cent success rates and are vastly superior to antibiotics, which only help about 25 per cent of people and caused the problem in the first place.[4] The procedure is surprisingly safe, and over 5,000 transplants have been performed.

Can a change of microbes cure obesity?

Although only obese mice have been found to clearly benefit at present, one small human pilot study of nine obese subjects receiving samples from lean donors found no significant change in weight but did see an improvement in blood glucose, insulin profiles and butyrate-producing microbes.[5] It's just a question of time until stool transplants are used to treat obesity in humans. One thirty-two-year-old lady from Rhode Island who was cured by a transplant of her recurrent C. diff infection caused by too many antibiotics was extremely grateful to be able to stop going to the bathroom twenty times per day. She had never had weight problems and her weight was steady before her infections. But sixteen months later she returned to the clinic complaining of weight gain. She had increased from an average weight of 62 kg to an obese 80 kg with a BMI of 34. A further ten months of monitored dieting failed to shift the weight. The lady had picked her own sixteen-year-old daughter as her preferred donor. At the time she was healthy and only slightly overweight at 63 kg but teenagers can change quickly and she rapidly gained 14 kg in the next two years and became obese. Another lady who gained weight after transplant from an anonymous donor was reported from New Orleans.[6] While these unfortunate cases are rare and other explanations are possible, it shows the potential of microbes to cause or cure obesity not just in mice but in humans – and the message is clear: choose your donors wisely.

Perhaps this will start for ex-athletes and supermodels a lucrative new career as donors, although you may want to avoid the anorectic microbes. Most people find the idea of a stool transplant, even if it is technically called an FMT (faecal microbiome transplant), too extreme. So if you are overweight and unhealthy but this is not for you, bear in mind that there is always bariatric surgery (gastric bypass), a form of auto-transplant that dramatically shifts the microbiota from one site to another. This appears to be permanent based on follow-up studies lasting nine years.

But if you don't fancy either of these procedures, what else can you do? How can you use these new insights into diet, microbes and food to help you lose weight?

If you want to lose a few pounds, avoid bursts of unsustainable dieting on restrictive food plans that inevitably rebound. The key, in fact, lies in not starting a classic diet in the first place, but in maintaining minor weight loss in a nutritionally sensible and sustainable way by remaking your whole diet as regards not just amounts but composition, variety and timing. Once again we need to remind ourselves that we and our microbes are all very individual and we must find the method that best suits us, not some inflexible formula. One bonus is that if you can lose weight even briefly and without major rebounds, it can lower your heart disease risk over your whole life.[7] Rapid weight loss through intermittent fasting looks as if it may be more likely to alter your microbial community for the better. The trick is to maintain it.

The thirty-ingredient heavy petting diet

'Knowing about my microbiome has totally changed my diet and what I put in my mouth.' Karen is a thirty-seven-year-old single mum who works in research in London. 'I became interested in microbiome research because of my weight problems and IBS since the age of fourteen. This got worse when I rapidly put on nearly five stone in a few years after having my daughter and giving up karate. I had tried lots of traditional diets and none of them had worked for me, so I was keen to try something different. I found out about this new DIY microbiome diet from the internet. The idea was for sixty days to eat as many different fruits, pulses [legumes] and vegetables [raw and unwashed, if possible] – ideally over thirty types per week, while petting a different animal per day. You could eat any other real [non-processed] foods you wanted, although grains were discouraged. It was far from a normal diet.

'I lost three kilos easily, although the exotic vegetable shopping was harder and more expensive than I had thought. Chasing squirrels to pet was also tricky. I continued with a less intensive regime of twenty items per week for another six months. I lost another three kilos in the next three months before putting most of it back on. The bad news was, after nearly a whole year my weight hadn't shifted dramatically – although the good news was my guts for the first time in fifteen years seemed normal. I went to the loo only once a day, not ten times, which was fantastic.'

Although feeling healthier, Karen repeated her varied prebiotic twenty-item diet, but this time after a kick-start by clearing out her colon with laxatives. She lost another five kilograms, and I then suggested longer-term intermittent fasting as a way of changing her microbes and losing weight. After three months, she lost a further five kilos, so she is now ten kilos lighter and, more importantly, she feels a lot better. We don't know whether the effects of the intermittent fasting would be permanent on her microbiome if she stopped, but we do see major changes in the species during fasting periods when mucus-eating microbes start their cleanup operations.[8]

Karen's story, although just one anecdote and far from a trial or proof, illustrates several points. Simply being aware of the new parallel universe living inside you helps the way you deal with food at a psychological level. As Liping Zhao the Chinese obesity specialist told me, 'My dieting patients in Shanghai really respond to knowing about their microbes and how they affect your health. It helps them comply with changes in their diet without being obsessed with weight scales.' It is also important not only to focus on body fat but to try to improve overall health first. So Karen's bowel disturbance, IBS, probably needed sorting before she could deal with the longer-term problem of fat, and the prebiotic diet appeared to do that very successfully.

People who are basically fit but want to be healthier can achieve this in a fairly straightforward way by helping their microbes in the ways I've suggested. Try to eat a greater variety of foods, particularly fruits, olive oil, nuts, vegetables and pulses plus fibre and polyphenols. Avoid processed foods, anything that claims to be a special low-fat or light product or has too many ingredients, and reduce your meat intake. Eat traditional cheese and full-fat yoghurt, avoiding synthetic varieties. Try adding more variety of fermented foods to your diet like kefir, fermented cabbage or soy-based foods. I like the concept that our ancestors ate in very irregular and seasonal ways, so intermittent fasting or giving up meat for months at a time or skipping some meals seems sensible to enhance your perception of variety. Throughout the year try to eat fruits and vegetables that are in season so as to increase the diversity of the foods you consume. Also, cutting back on liquid calories, such as sugar in juices and other drinks, as well as calories in cakes and snacks, is sensible, as is avoiding artificial sweeteners as a regular alternative.

Can poison be good for you?

Friedrich Nietzsche said: 'What does not kill me makes me stronger.' The data suggests our gut microbes like variety and the occasional shake-up, so could the odd drop of poison do us some good? Many believe that little doses of arsenic can be helpful. (Don't try this at home.) The concept of low doses of bad things being good for the system is called 'hormesis', from the Greek meaning 'to set in motion'. This idea is taken way too far, in my opinion, by homeopathy gurus who claim that dilutions of less than a molecule can have important biological effects – akin to peeing in the Atlantic Ocean and claiming to notice the difference.

Nevertheless, in much of biology, from the cell to the whole body, low levels of stress are good for the organism. Take, for example, short bursts of oxidants or heat stress, which make worms live longer, or low-dose antibiotics which make microbes much stronger, and too-low doses of anti-cancer drugs, which make cancer cells more resistant.[9] Even exercise is a form of stress, and we know that's good for us. In the same way, intermittent fasting can make small animals live longer – even fasting overnight or skipping breakfast may offer us a form of beneficial hormesis.

So we have to keep an open mind. Indulging in a once-a-year junk food blow-out or a greasy fry-up for breakfast could, counter-intuitively, keep us on our toes while fine-tuning our microbes and immune systems. Similarly, a once-a-year steak for a vegetarian or an exotic salad for a meat eater could do wonders. It is also a good excuse to lose a bit of control, but don't forget this only works if it comes in the form of an occasional shock, not a daily or hourly one, and on the foundation of a high-fibre, microbe-friendly diet.

The hormesis concept is helpful even if we don't yet understand how it works, and if used sparingly it contributes to the general aim of diet diversity. Every gardener knows that it can be hard to predict which plants will grow best in which soils without some trial and error. Given that our natural soils vary so much, self-experimentation and an open mind may offer us another useful route to health.

I met an entrepreneurial Goan scientist called Darryl on a plane. He gained a lot of weight when he moved from Oxford to New York and had bad experiences with expensive diet gurus. Since then he has taken a DIY approach. He obsessively takes photographs of every meal he eats on his iPad, and using the Timeline program links this to a diary of how he feels. When he tried a high-protein low-carb diet of kefir (like yoghurt and full of probiotics) and meat twice a day, his waist size dropped by two inches in six weeks and he felt energised and didn't need sleep. But the energy spilled over into aggression, and for the first time in his adult life he got into a fight. He realised high protein was not for him. He experimented with a fruitarian diet comprising little veg and plenty of fruit and coconuts, but found his weight increased. He is still experimenting with other variations, but seems to be settling into a low-refined-carbohydrate, high-vegetable and -legume Mediterranean-style diet. He enjoys the challenge of trying to increase his microbial diversity each week, and is happy with his weight.

Darryl gives us a glimpse of the future. As well as our genes, our microbes will soon be tested routinely from birth, as the cost of carrying out these tests will soon drop below routine blood tests. As technology becomes ever better and cheaper we can envisage more and more innovations and more personalised diet prescriptions. There are already mobile apps like Magic Snap, which can roughly estimate calories from photos of the food on your plate, and which will improve. These apps (including one we are developing) could be easily modified to provide microbe-friendly prebiotic food scores that we could combine with genetic read-outs of our daily microbes, providing a truly personal nutritional advice service.

Looking after your microbial garden

Most of you – the interested people reading this book – can make informed decisions about changing your dietary habits or not, about eating more of one food or less of another – or you could ignore me and your microbes altogether. However, many people because of limited resources or education cannot. But we are all responsible for the global sickness and obesity epidemic and must exert pressure on governments to start reversing the situation. Let's try to reduce the massive taxpayer subsidies for corn, soy and sugar which all go into processed foods, and make it worthwhile for farmers to grow cheaper fruits and vegetables.

There are other global actions that could be taken to improve our microbes and our health. We need more incentives to reduce the massive abuse of antibiotics, particularly in vulnerable children. We need to reduce the number of C-sections and focus on normal deliveries.

Furthermore, 'hygiene' is something we should redefine and bother less about – it used to be a term for avoiding excreting in the open, but has turned into an obsession with eliminating all natural smells and microbes from our bodies and especially our mouths. Our homes have become sterile laboratories, modern kitchens more like operating theatres, and food comes smothered in plastic. Children should be allowed to play outside in the dirt and swap their microbes with playmates and animals as much as possible. Perhaps we shouldn't over-wash our foods, and this could be a possible benefit of organic products. One study has shown that you ingest many more healthy live microbes when you have a varied diet with fresh fruit and vegetables, and this could prove to be healthy.[10] We should also rethink our debate about GM crops and focus more on whether the resulting changes to soil and plant environments could be harmful to our gut microbes. Gardeners are reported as being on average healthier and less depressed than the rest of us, and it could be that it's the close contact with dirt and microbes that makes the difference.

There is no such thing as a one-size-fits-all diet. A recurring theme of this book is that our guts and brains are so individual, and the ways we react to foods so different and yet flexible. Our lives can be a voyage of discovery to find out what works best for us. Having overturned, I hope, many diet and food myths, I also hope that you are now more sceptical about claims concerning food and diet, however convincingly they are sold to you. No one is the infallible expert or completely impartial in this massive field, where we find ten thousand times more recycled theories than rigorous experiments. We can create synthetic DNA and clone animals, but we still know incredibly little about the stuff that keeps us alive.

This book is about dispelling diet myths and arbitrary rules. I have tried not to replace them with new rules or restrictions, but rather with knowledge. You won't go wrong if you just treat your own microbes like you would treat your own garden. Give them plenty of fertiliser – prebiotics, fibre and nutrients. Plant new seeds regularly in the shape of probiotics and new foods. Give the soil an occasional rest by fasting. Experiment, but avoid poisoning your microbiotic garden with preservatives, antiseptic mouthwashes, antibiotics, junk food and sugar.

These treatments will maximise the diversity of the species that flourish, producing the greatest range of nutrients. In this way your personal garden will cope better with the occasional floods or droughts or invasions of toxic weeds – feasts and famines, infections and cancers. After riding the storm and inevitably suffering a few casualties, the range and balance of your gut flora will allow everything to regrow even more robustly, so allowing you and your microbes to stay healthy. Rather than thinking of your body as a temple, think of it as this precious garden.

Although we still have much to learn, my hunch is that diversity is the key.

Glossary

amylase: an enzyme produced in saliva and in the pancreas that breaks down starch in carbohydrates into glucose and energy. We all have different levels because of our genes.

antibiotics: chemicals originally produced by bacteria to defend themselves that have been manufactured for humans to use against bacteria. No new antibiotic drugs have been produced in thirty years and many bacteria have evolved resistance.

antioxidant: any beneficial substance that can mop up harmful products generated by cells like free radicals.

bacteria: simple but highly flexible tiny and ancient microbes that exist in every part of the world as well as in all the cavities of our bodies. Most are harmless, a few are disease-causing (pathogenic) and many are helpful.

Bacteroidetes: a common class (or phylum) of bacteria in our guts (Bacteroides are a sub-branch) that change with our environment and diet (e.g. meat eating).

bifidobacteria: subtypes of bacteria, commonly probiotics, that are contained in milk products and yoghurt as well as breast milk. Generally considered healthy in Western guts.

BMI (body mass index): a formula that estimates your amount of body fat by dividing your weight in kilograms by your height in metres squared. E.g. a 70 kg male 1.80 metres tall has a BMI of 70/3.24 (1.80 × 1.80) = 21.6. Above BMI 25 is overweight, and over 30 is classified as obese. BMI is not very accurate in distinguishing muscle from fat.

butyrate: a healthy short-chain fatty acid (SCFA) produced by bacteria when they digest food in the colon containing fibre and carbs, especially polyphenols. It has antioxidant and anti-inflammatory properties and signals the immune system to act.

carnitine: made inside the body from amino acids, and important in helping to fuel the body. When digested it can raise levels of TMAO and increase the risk of heart disease. Used as a supplement by bodybuilders to burn fat and increase muscle, and is found in large amounts in meat.

cholesterol: a class of essential lipid that the body synthesises itself to keep our cells intact. It is transported around the body by lipoproteins. High blood levels of cholesterol roughly correlate with heart disease, but its risk has been exaggerated. Found in many foods, including fish and nuts. Healthy levels are regarded as below 5mmol/l but the average in the UK are around 6mmol/l.

Christensenella: an ancient bacteria that a minority of people have in their guts that has been found to protect against obesity and visceral fat. Related to methane-producing bacteria.

Clostridium difficile (C. diff): a pathogenic bacteria that lives happily in the guts of 3 per cent of people until overdoses of antibiotics wipe out its competitors, at which point it grows rapidly and takes over in massive numbers. It can produce nasty toxins causing major gut damage (colitis). Often resistant to antibiotics, which just make it stronger. Often the only cure is a faecal microbial transplant (FMT).

colitis: inflammation of the colon caused by infections or autoimmune disease.

colon: the lower part of the intestines, where most of our bacteria and microbes live and digest the fibre-rich food that hasn't been absorbed higher up in the small intestine.

diabetes: two diseases resulting from too much sugar (glucose) in the blood. The commonest is Type 2, which is related to obesity and to our genes and makes insulin ineffective, causing blood glucose to rise and excessive insulin to be produced in compensation.

DNA: Deoxyribonucleic acid is the building block of our genetic material; it is arranged as a double helix in 23 chromosomes and contains the roughly 20,000 genes in each cell of our bodies.

E. coli: a common bacteria that lives in our colons and can occasionally become pathogenic after infections or antibiotics.

endocrine: a term for anything producing hormones (e.g. thyroid or pancreas).

endocrine disruptors: chemicals that act epigenetically to alter hormones, e.g. the bisphenol (BPA) in plastic bottles.

epidemiology: the study of large groups or populations in order to discover the causes of disease.

epigenetic: describes mechanisms by which chemical signals can switch genes on and off without altering the DNA structure. A normal process in babies and growth, which can be altered by diet and chemicals for up to several generations.

fat: a term with many different meanings. Scientifically synonymous with lipids.

fermentation: a process involving microbes that modifies and preserves food to create alcohol in beer or wine, or lactic acid in milk products or pickled cabbage.

fibre: a term for the hard-to-digest parts of carbohydrates that reach the colon for our microbes to feed off. High levels are found in fruit, legumes, other vegetables, whole grains and nuts. Artificial fibre is also available as additives.

Firmicutes: a major group (phylum) of our gut microbes that contains many of the species related to health and disease and is partly influenced by our genes.

FMT: a polite term for faecal microbial transplant, whereby a healthy donor's stool is placed via tubes or tablets into the colon of the recipient.

free radicals: small chemicals released from cells as part of their normal function. If they accumulate they can be harmful to the body. Mopped up by antioxidants.

fructose: a carbohydrate sugar that makes up 50 per cent of table sugar and is much sweeter. Contained in most fruits and can be produced artificially from corn syrup and used in soft drinks.

fungi: a large group (kingdom) of ancient organisms that include yeast, mould and mushrooms.

gene: a small group of chemicals on our DNA that tell the body to make a particular protein. We have about 20,000 in each of our cells. Estimates vary as our precise definition of genes changes.

glycaemic index (GI): a measure of the speed at which different foods produce an increase in glucose and then insulin in the blood. Low GI foods are the basis of many diets. High-GI foods enable sugars to be released rapidly and cause rapid peaks in blood glucose and insulin (e.g. mashed potatoes) as opposed to low-GI high-fibre foods like celery. It is still unclear how important this mechanism is for influencing obesity compared to microbes.

HDL (high-density lipoprotein): a combination of a lipid and a protein that safely transports fats around the body. Measured in the blood to compare in a ratio with the unhealthy form, low-density lipoproteins (LDL).

heritability: the proportion of the variation in a trait due to genetic effects or, more precisely, the differences between people in a trait or disease that can be explained by genetic factors. Ranges from 0 to 100 per cent.

hypothalamus: an area at the base of the brain controlling the release of many hormones including those relating to emotion, stress and appetite.

IBS (irritable bowel syndrome): a common complaint with no commonly agreed cause. It produces altered bowel frequency, cramps and wind. Associated with abnormal gut microbes and may be multiple conditions, some of which respond to specific fibre diets.

IGF-1: insulin growth factor 1 is a hormone implicated in many body functions, including the speed of ageing and repair. Has good evidence of influencing longevity in animals but not yet in humans.

inflammation: results from the way the body reacts to injury, infection or stress and is a normal response involving many different processes (e.g. after a bee sting). It usually causes leakiness of cell walls and the triggering of repair and defence cells, resulting in swelling, redness, pain, heat and loss of function.

insulin: the hormone which, in response to blood glucose, controls how much sugar to store as glycogen in the liver and fat in fat cells.

insulin resistance: results when insulin doesn't rise as much after ingesting glucose, forcing the pancreas to produce more insulin so as to control glucose levels; leads to diabetes.

inulin: a chemical that acts as a strong prebiotic helping bacteria to multiply. Found in high concentration in artichokes, chicory, garlic, onions, and in small amounts in bread.

ketogenic diet: a diet that makes the body change from burning glucose as energy to burning ketones, which derive from protein and fat. Examples include high-protein/low-carb diets and fasting.

lactobacillus: a bacteria which breaks down lactose in milk and in other sugars into lactic acid. A key component in many foods, changing their acidity and allowing them to be preserved, e.g. cheese, yoghurt and pickles.

leptin: a hormone released by the brain and closely related to body fat levels.

lipids: the scientific word for fats, but include many other molecules, e.g. fatty acids. When combined with proteins lipids are called lipoproteins, which travel around the body and can be of different shapes and sizes.

low-density lipoproteins (LDL): the less healthy form in which lipids are transported. They can get absorbed into blood vessels and lead to clogging of the arteries (atherosclerosis).

low-fat products: may just mean slighly less fat than normal, or that the fat has been replaced with sugars or proteins like soy and a multitude of chemicals to make them palatable.

medium-chain triglycerides: contain medium-chain fatty acids that are different from human or cow's milk. Are more ketogenic than other fats. Despite lack of evidence, believed by some to have health properties. Contained in palm and coconut oil.

meta-analysis: a technique for combining the results of different studies or trials to produce a single summary result. Provides better evidence than any single study, but can still be misleading if all the studies are biased.

metabolic rate: a measure of how rapid or slow energy input and output processes occur.

metabolism: the way the body and all cells both use and expend energy. Can be modified by many factors, e.g. heat, exercise, illness.

metabolomics: the study of the body's metabolites, which are the chemical signatures of the cell. Believed to number around 3,000 in human blood.

microbe: an organism that can be seen only with the aid of a microscope; includes bacteria, viruses, yeast and some larvae and worms.

microbiome: the whole community of microbes that may be in our guts or mouths or in the soil.

neurotransmitters: chemicals in the brain that allow nerve cells (neurons) to communicate and control mood (e.g. serotonin, dopamine).

observational study: a type of epidemiological research that makes inferences by comparing risk factors (e.g. food) with outcomes like disease. Evidence is weak when cross-sectional but better when people are followed over long periods (prospective observational study, or cohort study). All observational studies can be biased.

oil: a type of lipid (fat) that is liquid at room temperature.

oleic acid: a fatty acid (mono-unsaturated) that is one of the major components of olive oil.

omega 3 fatty acid: also called ω-3, this is a polyunsaturated fatty acid with a double bond at the 3rd carbon. Contained in many oily fish and often used as a (much hyped) health supplement for the heart and brain. It is an essential fatty acid that we can't make ourselves.

omega 6 fatty acid: also called ω-6, this is a polyunsaturated fatty acid with a double bond at the 6th carbon. Contained in many foods e.g. soybean, palm oil, chicken, nuts and seeds, and is an essential fatty acid. Has a bad (unfounded) reputation.

polyphenols: a group of many chemicals released from food after digestion by microbes, many of which are useful and healthy. Polyphenols include flavonoids and resveratrol, which have anti-oxidant properties. Contained in vegetables, fruits, nuts, tea, coffee, chocolate, beer and wine.

polyunsaturated fatty acids (PUFA): a type of lipid that consists of long-chain fatty acids with double bonds, and that are part of many foods generally regarded as healthy.

prebiotic: any food component that encourages healthy bacteria to flourish like a fertiliser. Found in all breast milk. Bacteria often feed off prebiotics. Often contains inulin, which is found in high levels in Jerusalem and globe artichokes, celery, garlic, onions and chicory root.

PREDIMED study: a Spanish clinical study which random-ly allocated over 7,000 patients to a low fat diet or a classical

Mediterranean diet and followed them for four years. They found the Mediterranean diet was superior in reducing heart disease, diabetes and weight gain.

Prevotella: a group of bacteria that flourishes in vegetarians and is rare in meat eaters. Generally regarded as a marker of a healthy diet.

probiotic: a group of bacteria added as a food supplement. Can be freeze-dried or found in fermented milk products (yoghurt, kefir), cabbage (sauerkraut, kimchi), tea (kombucha) or soy (miso). Believed to provide health benefits.

PROP (6-N-Propylthiouracil): a chemical tasting either very bitter or tasteless, depending on your genetic make-up. Used to test taste in experiments.

randomised controlled trial: an RCT is the gold standard for evidence in epidemiology. Subjects are randomly allocated to the test treatment or diet and compared to another standard treatment or a placebo, then followed for months or years (cf. PREDIMED study).

resveratrol: a type of polyphenol naturally found in food and wine that in animals has healthy anti-ageing properties. Large amounts need to be ingested to be beneficial; side effects from overdose have been reported.

saturated fat: a type of lipid that lacks hydrogen bonds and is contained in large amounts in oils like coconut and palm oil, dairy and meats. Previously thought to be harmful.

sequencing: a term for identifying all the key parts of DNA and genes in an organism. Usually the DNA is broken into millions of small pieces and reassembled (often referred to as 'shotgun'). Used to identify microbe species and disease genes in humans in great detail.

sugar: another common word for soluble carbohydrates; or refers to the white powder we eat called sucrose, which is a mix of glucose and fructose. The suffix -ose means that the chemical is a sugar (e.g. lactose).

TASR1 and TASR2: the major taste-receptor genes that are expressed by taste receptors found all over the oral cavity and that influence how we perceive sweet and bitter tastes.

TMA and TMAO: trimethylamine is a chemical in meat and large fish that is converted in the gut by microbes into the oxidised form, TMAO, which accelerates atherosclerosis and heart disease.

trans fats: also called hydrogenated fats, these are chemically transformed unsaturated fats that are easy to cook with but hard to break down in the body. Common as dairy substitutes and in junk foods. Major cause of heart disease and cancer. Banned in some countries and slowly being phased out in others.

Treg cell: an important type of immune cell (T-suppressor regulatory cell) that keeps the immune response in check and has two-way communication with gut microbes.

umami: the fifth (savoury) taste sense that mimics meat and comes from glutamate and is found in mushrooms. There is now a possible sixth sense called 'kokumi', meaning 'heartiness'.

viruses: tiny microbes that outnumber bacteria 5 to 1, and many feed off them (these devourers are called 'phages') to control their numbers. Most are harmless to us and live in our bodies and may have a health role.

visceral fat: the internal fat that accumulates around your intestines and liver. Excess fat is associated with heart disease and diabetes. More harmful than fat on the exterior of the body.

vitamins: molecules that are essential for the body's chemical reactions to work. We get most from food, sunlight (vitamin D) and our gut microbes.

yeast: a member of the fungi kingdom that converts sugar into alcohol and carbon dioxide. Used in making bread and alcohol. May promote healthy gut microbes. Can happily live in our guts and is only rarely pathogenic, e.g. candida infections.

Acknowledgements

This book has been a long time in gestation and many people have helped me along the way to deliver it, whether by a chance comment or a detailed explanation or sending me wacky YouTube clips. Getting the book off the ground is often the hardest part, and my agent and friend Sophie Lambert of Conville & Walsh has been essential and a fantastic support. The book has been through several transformations, each one guided enthusiastically by my highly skilled editor at Weidenfeld & Nicolson, Bea Hemming, who I learnt from and who was a pleasure to work with. Several people deserve special mention. Kirsten Ward helped me research many of the topics and select the case histories, assisted by my intuitive assistant Victoria Vazquez, both of whom diligently read some very early bad drafts. Later versions were read and improved by old and trusted friends Robyn Fitzgerald, Roz Kadir, Bryan Fehilly and Frannie Hochberg.

My colleagues in obesity genetics, nutrition and the microbiome were invaluable sources of the latest information before it was published. These include my microbe collaborators Ruth Ley and Rob Knight, plus his team from American Gut, Dan McDonald and Luke Thompson; Dusko Ehrlich, Peter Turnbaugh, Paul O'Toole, Glen Gibson, Susan Erdman, Stan Hazen, Amir Zarrinpar, Marty Blaser, Maria Dominguez-Bello, Patrice Cani, Kevin Tuohy, Laurel Lagenaur, Rashmi Sinha and Jim Goedert, as well as my King's College London team, Michelle Beaumont, Jordana Bell, Craig Glastonbury and Matt Jackson. Other brains I quizzed include Steve O'Rahilly, George Davy Smith, Mark McCarthy, David Allison, Clare Llewellyn, Kirsi Pietiläinen, Ele Zeggini, Alina Farmaki, Barbara Prainsack, Aubrey Sheiham, Leora Eisen, Stephen O'Hara and David Morgan.

I had very helpful conversations with my KCL colleagues Kevin Whelan, Jeremy Sanderson, Phil Chowienczyk and Tom Sanders.

In Barcelona, merci to my generous host Xavier Estivill and Ramón Estruch, Mark Nieuwenhuijsen, Susana Puig and Josep Malvehy for their knowledge and hospitality. I learnt about cheese from Paul Neale, Jon Schofield, Nigel White and Eric Biseaux, and about food philosophy from Julian Baggini. Helpful tips came from Ian Weir, Aine Kelly, Isgar Boss, Peter Kindersley, Michael Antoniou, Amanda Bailey, John Hemming, Swami, Brenda Sambrook, Lesley Bookbinder, Vivienne Hall and others who wrote books on food and diet which I dipped into, or whose names I have forgotten but whose tips were equally helpful. I thank my gifted son Thomas for his junk-food sacrifices and the other diligent and adventurous dieters, and all the TwinsUK volunteers. I must thank St Thomas' Hospital, King's College London, my funders at the EU, NIHR and Wellcome Trust and as always the amazing twins for providing me with such a perfect base for research, and Veronique, Sophie and Tom for generously allowing me to disappear from real life for long periods and occupy the freezer with strange samples.

Finally, despite all the help I received, this is a vast and fast-moving area of science which means I will surely have made some errors and omissions, for which I happily take any blame.

Notes

Introduction

1 Imamura, F., *The Lancet Global Health* (2015); 3(3): e132 DOI: 10.1016/S2214-109X(14)70381-X. Dietary quality among men and women in 187 countries in 1990 and 2010: a systematic assessment.

2 Pietiläinen, K.H., *Int J Obes* (Mar 2012); 36(3): 456–64. Does dieting make you fat? A twin study.

3 Ochner, C.N., *Lancet Diabetes Endocrinol* (11 Feb 2015); pii: S2213-8587(15)00009-1. doi: 10.1016/S2213-8587(15)00009-1. Treating obesity seriously: when recommendations for lifestyle change confront biological adaptations.

4 Goldacre, B., *Bad Science* (Fourth Estate, 2008); www.quackwatch.com/04ConsumerEducation/nutritionist.html

1 Not on the Label: Microbes

1 Fernández, L., *Pharmacol Res* (Mar 2013); 69(1): 1–10. The human milk microbiota: origin and potential roles in health and disease.

2 Aagaard, K., *Sci Transl Med* (21 May 2014); 6(237): 237ra65. The placenta harbors a unique microbiome.

3 Funkhouser, L.J., *PLoS Biol* (2013); 11(8): e1001631. Mom knows best: the universality of maternal microbial transmission.

4 Koren, O., *Cell* (3 Aug 2012); 150(3): 470–80. Host remodeling of the gut microbiome and metabolic changes during pregnancy.

5 Hansen, C.H., *Gut Microbes* (May–Jun 2013); 4(3): 241–5. Customizing laboratory mice by modifying gut microbiota and host immunity in an early 'window of opportunity'.

6 http://www.britishgut.org and http://www.americangut.org

7 Afshinnekoo, E., *CELS* (2015); http://dx.doi.org/10.1016/j.cels.2015.01.0012015. Geospatial resolution of human and bacterial diversity with city-scale metagenomics.

2 Energy and Calories

1 Kavanagh, K., *Obesity* (Jul 2007); 15(7): 1675–84. Trans fat diet induces abdominal obesity and changes in insulin sensitivity in monkeys.

2 Novotny, J.A., *American Journal of Clinical Nutrition* (1 Aug 2012); 96(2): 296–301. Discrepancy between the Atwater Factor predicted and

empirically measured energy values of almonds in human diets.

3 Bleich, S.N., *Am J Prev Med* (6 Oct 2014); pii: S07493797(14)00493-0. Calorie changes in chain restaurant menu items: implications for obesity and evaluations of menu labelling.

4 Sun, L., *Physiol Behav* (Feb 2015); 139: 505–10. The impact of eating methods on eating rate and glycemic response in healthy adults.

5 Sacks, F.M., *JAMA* (17 Dec 2014); 312: 2531–41. Effects of high vs low GI of dietary carbohydrate and insulin sensitivity: the OmniCarb RCT.

6 Zeevi, D., *Cell* (19 Nov 2015); 163/5): 1079–94. Personalized nutrition by prediction of glycemic responses.

7 Bouchard, C., *N Engl J Med* (24 May 1990); 322(21): 1477–82. The response to long-term overfeeding in identical twins.

8 Samaras, K., *J Clin Endocrinol Metab* (Mar 1997); 82(3): 781–5. Independent genetic factors determine the amount and distribution of fat in women after the menopause.

9 Stubbe, J.H., *PLoS One* (20 Dec 2006); 1: e22. Genetic influences on exercise participation in 37,051 twin pairs from seven countries.

10 Neel, J.V., *Am J Hum Genet* (Dec 1962); 14: 353–62. Diabetes mellitus: a 'thrifty' genotype rendered detrimental by 'progress'?

11 Song, B., *J Math Biol* (2007) 54: 27–43. Dynamics of starvation in humans.

12 Speakman, J.R., *Int J Obes* (Nov 2008); 32(11): 1611–17. Thrifty genes for obesity: the 'drifty gene' hypothesis.

13 Speakman, J.R., *Physiology* (Mar 2014); 29(2): 88–98. If body fatness is under physiological regulation, then how come we have an obesity epidemic?

14 Mustelin, L., *J Appl Physiol* (1985) (Mar 2011); 110(3): 681–6. Associations between sports participation, cardiorespiratory fitness, and adiposity in young adult twins.

15 Ogden, C.L., *JAMA* (26 Feb 2014); 311(8): 806–14. Prevalence of childhood and adult obesity in the United States, 2011–12.

16 Rokholm, B., *Obes Rev* (Dec 2010); 11(12): 835–46. The levelling off of the obesity epidemic since the year 1999 – a review of evidence and perspectives.

17 Lee, R.J., *J Clin Invest* (3 Mar 2014); 124(3): 1393–405. Bitter and sweet taste receptors regulate human upper respiratory innate immunity.

18 Negri, R., *J Pediatr Gastroenterol Nutr* (May 2012); 54(5): 624–9. Taste perception and food choices.

19 Keskitalo, K., *Am J Clin Nutr* (Aug 2008); 88(2): 263–71. The Three-factor Eating Questionnaire, body mass index, and responses to sweet and salty fatty foods: a twin study of genetic and environmental associations.

20 Fushan, A.A., *Curr Biol* (11 Aug 2009); 19(15): 1288–9. Allelic

polymorphism within the TAS1R3 promoter is associated with human taste sensitivity to sucrose.

21 Mennella, J.A., *PLoS One* (2014); 9(3): e92201. Preferences for salty and sweet tastes are elevated and related to each other during childhood.

22 Mosley, M., *Fast Exercise* (Atria Books, 2013)

23 Stubbe, J.H., *PLoS One* (20 Dec 2006); 1: e22. Genetic influences on exercise participation in 37,051 twin pairs from seven countries.

24 den Hoed, M., *Am J Clin Nutr* (Nov 2013); 98(5): 1317–25. Heritability of objectively assessed daily physical activity and sedentary behavior.

25 Archer, E., *Mayo Clin Proc* (Dec 2013); 88(12): 1368–77. Maternal inactivity: 45-year trends in mothers' use of time.

26 Gast, G-C. M., *Int J Obes* (2007); 31: 515–20. Intra-national variation in trends in overweight and leisure time physical activities in the Netherlands since 1980: stratification according to sex, age and urbanisation degree.

27 Westerterp, K.R., *Int J Obes* (Aug 2008); 32(8): 1256–63. Physical activity energy expenditure has not declined since the 1980s and matches energy expenditures of wild mammals.

28 Spector, T.D., *BMJ* (5 May 1990); 300(6733): 1173–4. Trends in admissions for hip fracture in England and Wales, 1968–85.

29 Hall, K.D., *Lancet* (27 Aug 2011); 378(9793): 826–37. Quantification of the effect of energy imbalance on bodyweight.

30 Williams, P.T., *Int J Obes* (Mar 2006); 30(3): 543–51. The effects of changing exercise levels on weight and age-related weight gain.

31 Hall, K.D., *Lancet* (27 Aug 2011); 378(9793): 826–37. Quantification of the effect of energy imbalance on bodyweight.

32 Turner, J.E., *Am J Clin Nutr* (Nov 2010); 92(5): 1009–16. Nonprescribed physical activity energy expenditure is maintained with structured exercise and implicates a compensatory increase in energy intake.

33 Strasser, B., *Ann NY Acad Sci* (Apr 2013); 1281: 141–59. Physical activity in obesity and metabolic syndrome.

34 Dombrowski, S.U., *BMJ* (14 May 2014); 348: g2646. Long-term maintenance of weight loss with non-surgical interventions in obese adults: systematic review and meta-analyses of randomised controlled trials.

35 Ekelund, U., *Am J Clin Nutr* (14 Jan 2015). Activity and all-cause mortality across levels of overall and abdominal adiposity in European men and women: the European Prospective Investigation into Cancer and Nutrition Study.

36 Hainer, V., *Diabetes Care* (Nov 2009); 32 Suppl 2: S392. Fat or fit: what is more important?

37 Fogelholm, M., *Obes Rev* (Mar 2010); 11(3): 202–21. Physical activity, fitness and fatness: relations to mortality, morbidity and disease risk factors.

38 Viloria, M., *Immunol Invest* (2011); 40: 640–56. Effect of moderate exercise on IgA levels and lymphocyte count in mouse intestine.

39 Matsumoto, M., *Biosci, Biotechnol Biochem* (2008); 72: 572–6. Voluntary running exercise alters microbiota composition and increases n-butyrate concentration in the rat cecum.

40 Hsu, Y.J., *J Strength Cond Res* (20 Aug 2014). Effect of intestinal microbiota on exercise performance in mice.

41 Clarke, S.F., *Gut* (Dec 2014); 63(12): 1913–20. Exercise and associated dietary extremes impact on gut microbial diversity.

42 Kubera, B., *Front Neuroenergetics* (8 Mar 2012); 4: 4. The brain's supply and demand in obesity.

3 Fats: Total

1 de Nijs, T., *Crit Rev Clin Lab Sci* (Nov 2013); 50(6): 163–71. ApoB versus non-HDL-cholesterol: diagnosis and cardiovascular risk management.

2 Kaur, N., *J Food Sci Technol* (Oct 2014); 51(10): 2289–303. Essential fatty acids as functional components of foods, a review.

3 Chowdhury, R., *Ann Intern Med* (18 Mar 2014); 160(6): 398–406. Association of dietary, circulating, and supplement fatty acids with coronary risk: a systematic review and meta-analysis.

4 Würtz, P., *Circulation* (8 Jan 2015). pii:114.013116. Metabolite profiling and cardiovascular event risk: a prospective study of three population-based cohorts.

5 Albert, B.B., *Sci Rep* (21 Jan 2015); 5: 7928. doi: 10.1038/srep07928. Fish oil supplements in New Zealand are highly oxidised and do not meet label content of n-3 PUFA.

6 Ackman, R.G., *J Am Oil Chem Soc* (1989); 66: 1162–64. EPA and DHA contents of encapsulated fish oil products.

7 Opperman, M., *Cardiovasc J Afr* (2011); 22: 324–29. Analysis of omega-3 fatty acid content of South African fish oil supplements.

8 Micha, R., *BMJ* (2014); 348: g2272. Global, regional, and national consumption levels of dietary fats and oils in 1990 and 2010: a systematic analysis including 266 country-specific nutrition surveys.

9 Campbell, T.C., *Am J Cardiol* (26 Nov 1998); 82(10B): 18T–21T. Diet, lifestyle, and the etiology of coronary artery disease: the Cornell China study.

10 Campbell, T.C., *The China Study* (BenBella Books, 2006)

4 Fats: Saturated

1 Law, M., *BMJ* (1999); 318: 1471–80. Why heart disease mortality is low in France: the time lag explanation.

2 Bertrand, X., *J Appl Microbiol* (Apr 2007); 102(4): 1052-9. Effect of cheese consumption on emergence of antimicrobial resistance in the intestinal

microflora induced by a short course of amoxicillin-clavulanic acid.

3   https://www.youtube.com/watch?v=134aMOQwyhY

4   David, L.A., *Nature* (23 Jan 2014); 505(7484): 559–63.

5   Teicholz, N., *The Big Fat Surprise* (Simon & Schuster, 2014)

6   Goldacre, B., *BMJ* (2014); 349 doi: http://dx.doi.org/10.1136/bmj.g4745. Mass treatment with statins.

7   Harborne, Z., *Open Heart* (2015); 2: doi:10.1136/openhrt-2014-000196. Evidence from randomised controlled trials did not support the introduction of dietary fat guidelines in 1977 and 1983: a systematic review and meta-analysis.

8   Siri-Tarino, P.W., *Am J Clin Nutr* (2010); 91: 535–46. Meta-analysis of prospective cohort studies evaluating the association of saturated fat with cardiovascular disease.

9   Hjerpsted, J., *Am J Clin Nutr* (2011); 94: 1479–84. Cheese intake in large amounts lowers LDL-cholesterol concentrations compared with butter intake of equal fat content.

10 Rice, B.H., *Curr Nutr Rep* (15 Mar 2014); 3: 130–38. Dairy and Cardiovascular Disease: A Review of Recent Observational Research.

11 Tachmazidou, I., *Nature Commun* (2013); 4: 2872. A rare functional cardioprotective APOC3 variant has risen in frequency in distinct population isolates.

12 Minger, D., *Death by Food Pyramid* (Primal Blueprint, 2013)

13 Chen, M., *Am J Clin Nutr* (Oct 2012); 96(4): 735–47. Effects of dairy intake on body weight and fat: a meta-analysis of randomized controlled trials.

14 Martinez-Gonzalez, M., *Nutr Metab Cardiovasc Dis* (Nov 2014); 24(11): 1189–96. Yogurt consumption, weight change and risk of overweight/ obesity: the SUN cohort.

15 Jacques, P., *Am J Clin Nutr* (May 2014); 99(5): 1229S–34S. Yogurt and weight management.

16 Kano, H., *J Dairy Sci* (2013); 96: 3525–34. Oral administration of Lactobacillus delbrueckii subspecies bulgaricus OLL1073R-1 suppresses inflammation by decreasing interleukin-6 responses in a murine model of atopic dermatitis.

17 Daneman, N., *Lancet* (12 Oct 2013); 382(9900): 1228–30 A probiotic trial: tipping the balance of evidence?

18 Guo, Z., *Nutr Metab Cardiovasc Dis* (2011); 21: 844–50. Influence of consumption of probiotics on the plasma lipid profile: a meta-analysis of randomised controlled trials.

19 Jones, M.L., *Br J Nutr* (May 2012); 107(10): 1505–13. Cholesterol-lowering efficacy of a microencapsulated bile salt hydrolase-active *Lactobacillus reuteri* NCIMB 30242 yoghurt formulation in hypercholesterolaemic adults.

20 Jones, M.L., *Eur J Clin Nutr* (2012); 66: 1234–41. Cholesterol lowering and inhibition of sterol absorption by Lactobacillus reuteri NCIMB 30242: a randomized controlled trial.

21 Morelli, L., *Am J Clin Nutr* (2014); 99(suppl): 1248S–50S. Yogurt, living cultures, and gut health.

22 McNulty, N.P., *Sci Transl Med* (26 Oct 2011); 3:106. The impact of a consortium of fermented milk strains on the gut microbiome of gnotobiotic mice and monozygotic twins.

23 Goodrich, J.K., *Cell* (6 Nov 2014); 159(4): 789–99. Human genetics shape the gut microbiome.

24 Roederer, M., *The Genetic Architecture of the Human Immune System* (*Cell*, 2015)

25 Saulnier, D.M., *Gut Microbes* (Jan–Feb 2013); 4(1): 17–27. The intestinal microbiome, probiotics and prebiotics in neurogastroenterology.

26 De Palma, G., *Gut Microbes* (May–Jun 2014); 5(3): 439–45. The microbiota-gut-brain axis in functional gastrointestinal disorders.

27 Sachdev, A.H., *Curr Gastroenterol Rep* (Oct 2012); 14(5): 439–45. Antibiotics for irritable bowel syndrome.

28 Idem n24.

29 Tillisch, K., *Gastroenterology* (Jun 2013); 144(7): 1394–401. Consumption of fermented milk product with probiotic modulates brain activity.

30 Tillisch, K., *Gut Microbes* (May–Jun 2014); 5(3): 404–10. The effects of gut microbiota on CNS function in humans.

5 Fats: Unsaturated

1 The rhyme could also relate to the unpopular taxes they raised, or even to an earlier dispute between King Richard and his brother John.

2 Daniel, C.R., *Public Health Nutr* (Apr 2011); 14(4): 575–83. Trends in meat consumption in the United States.

3 http://www.eblex.org.uk/wp/wp-content/uploads/2014/02/m_uk_yearbook13_Cattle110713.pdf

4 Micha, R., *Lipids* (Oct 2010); 45(10): 893–905. Saturated fat and cardiometabolic risk factors, coronary heart disease, stroke, and diabetes: a fresh look at the evidence.

5 Siri-Tarino, P.W., *Am J Clin Nutr* (2010); 91: 535–46. Meta-analysis of prospective cohort studies evaluating the association of saturated fat with cardiovascular disease.

6 Price, W.A., *Nutrition and Physical Degeneration*, 6th edn (La Mesa, Ca, Price-Pottenger Nutritional Foundation, 2003)

7 Willett, W.C., *Am J Clin Nutr* (Jun 1995); 61(6 Suppl): 1402S–6S. Mediterranean diet pyramid: a cultural model for healthy eating.

8 Estruch, R., *N Engl J Med* (4 Apr 2013); 368(14): 1279–90. Primary prevention of cardiovascular disease with a Mediterranean diet.

9 Salas-Salvadó, J., *Ann Intern Med* (7 Jan 2014); 160(1): 1–10. Prevention of diabetes with Mediterranean diets: a subgroup analysis of a randomized trial.

10 Guasch-Ferré, M., *BMC Med* (2014); 12: 78. Olive oil intake and risk of cardiovascular disease and mortality in the PREDIMED Study.

11 Konstantinidou, V., *FASEB J* (Jul 2010); 24(7): 2546–57. In vivo nutrigenomic effects of virgin olive oil polyphenols within the frame of the Mediterranean diet: a randomized controlled trial.

12 Lanter, B.B., MBio (2014); 5(3): e01206–14. Bacteria present in carotid arterial plaques are found as biofilm deposits which may contribute to enhanced risk of plaque rupture.

13 Vallverdú-Queralt, A., *Food Chem* (15 Dec 2013); 141(4): 3365–72. Bioactive compounds present in the Mediterranean sofrito.

6 Trans Fats

1 https://www.youtube.com/watch?v=zrv78nG9R04

2 Lam, H.M., *Lancet* (8 Jun 2013); 381(9882): 2044–53. Food supply and food safety issues in China.

3 Mozaffarian, D., *N Engl J Med* (2006); 354: 1601–13. Trans fatty acids and cardiovascular disease.

4 Kris-Etherton, P.M., *Lipids* (Oct 2012); 47(10): 931–40. Trans fatty acid intakes and food sources in the U.S. population: NHANES 1999–2002.

5 Thomas, L.H., *Am J Clin Nutr* (1981); 34: 877–86. Hydrogenated oils and fats: the presence of chemically-modified fatty acids in human adipose tissue.

6 Iqbal, M.P., *Pak J Med Sci* (Jan 2014); 30(1): 194–7. Trans fatty acids – a risk factor for cardiovascular disease.

7 Kishino, S., *Proc Natl Acad Sci* (29 Oct 2013); 110(44): 17808–13. Polyunsaturated fatty acid saturation by gut lactic acid bacteria affecting host lipid composition.

8 Pacifico, L., *World J Gastroenterol* (21 Jul 2014); 20(27): 9055–71. Nonalcoholic fatty liver disease and the heart in children and adolescents.

9 Mozaffarian, D., *N Engl J Med* (3 Jun 2011); 364(25): 2392–404. Changes in diet and lifestyle and long-term weight gain in women and men.

10 http://www.dailymail.co.uk/news/article-2313276/Man-keeps-McDonalds-burger-14-years-looks-exactly-the-day-flipped-Utah.html

11 Moss, M., *Salt, Sugar, Fat: How the Food Giants Hooked Us* (WH Allen, 2013)

12 Johnson, P.M., *Nat Neurosci* (May 2010);13(5): 635–41. Dopamine D2 receptors in addiction-like reward dysfunction and compulsive eating in obese rats.

13 Avena, N.M., *Methods Mol Biol* (2012); 829: 351–65. Animal models of sugar and fat bingeing: relationship to food addiction and increased body weight.

14 Taylor, V.H., *CMAJ* (9 Mar 2010);182(4): 327–8. The obesity epidemic: the role of addiction.

15 Bayol, S.A., *Br J Nutr* (Oct 2007); 98(4): 843–51. A maternal 'junk food' diet in pregnancy and lactation promotes an exacerbated taste for 'junk food' and a greater propensity for obesity in rat offspring.

16 David, L.A., *Nature* (23 Jan 2014); 505(7484): 559–61. Diet rapidly and reproducibly alters the human gut microbiome.

17 Chassaing, B., *Nature* (5 Mar 2015); 519: 92–6. Dietary emulsifiers impact the mouse gut microbiota promoting colitis and metabolic syndrome.

18 Poutahidis, T., *PLoS One* (Jul 2013); 10; 8(7): e68596. Microbial reprogramming inhibits Western diet-associated obesity.

19 Martinez-Medina, M., *Gut* (Jan 2014); 63(1): 116–24. Western diet induces dysbiosis with increased E coli in CEABAC10 mice, alters host barrier function favouring AIEC colonisation.

20 Huh, J.Y., *Mol Cells* (May 2014); 37(5): 365–71. Crosstalk between adipocytes and immune cells in adipose tissue inflammation and metabolic dysregulation in obesity.

21 Wang, J., *ISME J* (2015) 9: 1–15; Modulation of gut microbiota during probiotic-mediated attenuation of metabolic syndrome in high-fat-diet-fed mice.

22 Cox, A.J., *Lancet Diabetes Endocrinol* (Jul 2014); 21.pii: S2213–8587. Obesity, inflammation, and the gut microbiota.

23 Mraz, M., *J Endocrinol* (8 Jul 2014); pii: JOE-14-0283. The role of adipose tissue immune cells in obesity and low-grade inflammation.

24 Kong, L.C., *PLoS One* (Oct 2014); 20; 9(10): e109434. Dietary patterns differently associate with inflammation and gut microbiota in overweight and obese subjects.

25 Ridaura, V.K., *Science* (6 Sep 2013); 341(6150): 1241214. Gut microbiota from twins discordant for obesity modulate metabolism in mice.

26 Goodrich, J.K., *Cell*, (Nov 2014); 159(4): 789–99. Human genetics shape the gut microbiome.

27 Fei, N., *ISME J* (2013); 7: 880–4. An opportunistic pathogen isolated from the gut of an obese human causes obesity in germ-free mice.

28 Backhed, F., *Proc Natl Acad Sci USA* (2004); 101: 15718–15723. The gut microbiota as an environmental factor that regulates fat storage.

29 Backhed, F., *Proc Natl Acad Sci USA* (2007); 104: 979–84. Mechanisms underlying the resistance to diet-induced obesity in germ-free mice.

30 http://www.youtube.com/watch?v=gzL0fRkqjK8

31 Xiao, S., *FEMS Microbiol Ecol* (Feb 2014); 87(2): 357–67. A gut microbiota-targeted dietary intervention for amelioration of chronic inflammation underlying metabolic syndrome.

32 Zhou, H., *Dongjin Dynasty* (Tianjin Science & Technology Press, 2000)

33 Zhang, X., *PLoS One* (2012); 7(8): e42529. Structural changes of gut microbiota during berberine-mediated prevention of obesity and insulin resistance in high-fat diet-fed rats.

34 http://www.sciencemag.org/content/342/6162/1035

35 Alcock, J., *Bioessays* (8 Aug 2014); doi: 10.1002/bies.201400071. Is eating behavior manipulated by the gastrointestinal microbiota? Evolutionary pressures and potential mechanisms.

36 Vijay-Kumar, M., *Science* (9 Apr 2010); 328(5975): 228–31. Metabolic syndrome and altered gut microbiota in mice lacking Toll-like receptor 5.

37 Shin, S.C., *Science* (2011); 334 (6056): 670–4. Drosophila microbiome modulates host developmental and metabolic homeostasis via insulin signalling.

38 Tremaroli, V., *Nature* (13 Sep 2012); 489(7415): 242–9. Functional interactions between the gut microbiota and host metabolism.

7 Protein: Animal

1 Diamond, J., *Guns, Germs and Steel* (Norton, 1997)

2 Atkins, R., *The Diet Revolution* (Bantam Books, 1981)

3 Bueno, N.B., *Br J Nutr* (Oct 2013); 110(7): 1178–87. Very-low-carbohydrate ketogenic diet v. low-fat diet for long-term weight loss: a meta-analysis of randomised controlled trials.

4 Paoli, A., *Int J Environ Res Public Health* (19 Feb 2014); 11(2): 2092–107. Ketogenic diet for obesity: friend or foe?

5 Douketis, J.D., *Int J Obes* (2005); 29(10): 1153–67. Systematic review of long-term weight loss studies in obese adults: clinical significance and applicability to clinical practice.

6 Ebbeling, C.B., *JAMA* (27 Jun 2012); 307(24): 2627–34. Effects of dietary composition on energy expenditure during weight-loss maintenance.

7 Ibid.

8 Ellenbroek, J.H., *Am J Physiol Endocrinol Metab* (1 Mar 2014); 306(5): E552-8. Long-term ketogenic diet causes glucose intolerance and reduced beta and alpha cell mass but no weight loss in mice.

9 Cotillard, A., *Nature* (29 Aug 2013); 500(7464): 585–8. Dietary intervention impact on gut microbial gene richness.

10 Le Chatelier, E., *Nature* (29 Aug 2013); 500(7464): 541–6. Richness of human gut microbiome correlates with metabolic markers.

11 Qin, J., *Nature* (4 Oct 2012); 490(7418): 55–60. A metagenome-wide association study of gut microbiota in type 2 diabetes.

12 Fraser, G.E., *Arch Intern Med* (2001); 161: 1645–52. Ten years of life: is it a matter of choice?

13 Le, L.T., *Nutrients* (27 May 2014); 6(6): 2131–47. Beyond meatless, the health effects of vegan diets: findings from the Adventist cohorts.

14 Boomsma, D.I., *Twin Res* (Jun 1999); 2(2): 115–25. A religious upbringing reduces the influence of genetic factors on disinhibition: evidence for interaction between genotype and environment on personality.

15 Key, T.J., *Am J Clin Nutr* (2009); 89; 1613S–19S. Mortality in British vegetarians: results from the European Prospective Investigation into Cancer and Nutrition (EPIC Oxford).

16 Key, T.J., *Am J Clin Nutr* (Jun 2014); 100(Supplement 1): 378S–85S. Cancer in British vegetarians.

17 http://crossfitanaerobicinc.com/paleo-nutrition/list-of-foods/

18 Hidalgo, G., *Am J Hum Biol* (10 Sep 2014); 26(5): 710–12. The nutrition transition in the Venezuelan Amazonia: increased overweight and obesity with transculturation.

19 Schnorr, S.L., *Nat Commun* (15 Apr 2014); 5: 3654.doi: 10.1038/ncomms4654. Gut microbiome of the Hadza hunter-gatherers.

20 Dominguez-Bello, G., Personal communication.

21 Pan, A., *Arch Intern Med* (9 Apr 2012); 172(7): 555–63. Red meat consumption and mortality: results from 2 prospective cohort studies.

22 Rohrmann, S., *BMC Med* (7 Mar 2013); 11: 63.doi: 10.1186/1741-7015-11-6. Meat consumption and mortality – results from the European Prospective Investigation into Cancer and Nutrition.

23 Lee, J.E., *Am J Clin Nutr* (Oct 2013); 98(4): 1032–41. Meat intake and cause-specific mortality: a pooled analysis of Asian prospective cohort studies.

24 Tang, W.W., *N Engl J Med* (25 Apr 2013); 368(17): 1575–84. Intestinal microbial metabolism of phosphatidylcholine and cardiovascular risk.

25 Brown, J.M., *Curr Opin Lipidol* (Feb 2014); 25(1): 48–53. Metaorganismal nutrient metabolism as a basis of cardiovascular disease.

26 Kotwal, S., *Circ Cardiovasc Qual Outcomes* (2012); 5: 808–18. Omega 3 fatty acids and cardiovascular outcomes: systematic review and meta-analysis.

27 Mozaffarian, D., *JAMA* (2006); 296(15): 1885–99. Fish intake, contaminants, and human health: evaluating the risks and the benefits.

28 Lajous, M., *Am J Epidemiol* (1 Aug 2013); 178(3): 382–91. Changes in fish consumption in midlife and the risk of coronary heart disease in men and women.

29 http://www.dailymail.co.uk/health/article-2530164/Too-good-true-Diet-drink-acts-like-gastric-band-help-people-lose-stone-claim-scientists.html

30 Karimian, J., *J Res Med Sci* (Oct 2011); 16(10): 1347–53. Supplement consumption in bodybuilder athletes.

31 O'Dea, K., *Diabetes* (1984); 33: 596–603. Marked improvement in carbohydrate and lipid metabolism in diabetic Australian Aborigines after temporary reversion to traditional lifestyle.

32 http://www.theguardian.com/lifeandstyle/2013/may/12/type-2-diabetes-diet-cure

33 Look AHEAD Research Group, *N Engl J Med* (2013); 369: 145–54. Cardiovascular effects of intensive lifestyle intervention in Type 2 diabetes.

34 Franco, M., *BMJ* (9 Apr 2013); 346: f1515. Population-wide weight loss and regain in relation to diabetes burden and cardiovascular mortality in Cuba 1980–2010.

35 Mann, G.V., *J Atheroscler Res* (Jul–Aug 1964); 4: 289–312. Cardiovascular disease in the Masai.

8 Protein: Non-animal

1 Frankenfeld, C.L., *Am J Clin Nutr* (May 2011); 93(5): 1109–16. Dairy consumption is a significant correlate of urinary equol concentration in a representative sample of US adults.

2 Fritz, H., *PLoS One* (28 Nov 2013); 8(11): e81968. Soy, red clover and isoflavones, and breast cancer: a systematic review.

3 Lampe, J.W., *J Nutr* (Jul 2010); 140(7): 1369S–72S. Emerging research on equol and cancer.

4 Soni, M., *Maturitas* (Mar 2014); 77(3): 209–20. Phytoestrogens and cognitive function: a review.

5 Spector, T.D., *Lancet* (1982). Coffee, soya and cancer of the pancreas (letter).

6 Tsuchihashi, R., *J Nat Med* (Oct 2008); 62(4): 456–60. Microbial metabolism of soy isoflavones by human intestinal bacterial strains.

7 Renouf, M., *J Nutr* (Jun 2011); 141(6): 1120–6. *Bacteroides uniformis* is a putative bacterial species associated with the degradation of the isoflavone genistein in human feces.

8 Pudenz, M., *Nutrients* (15 Oct 2014); 6(10): 4218–72. Impact of soy isoflavones on the epigenome in cancer prevention.

9 Hehemann, J.H., *Nature* (8 Apr 2010); 464(7290): 908–12. Transfer of carbohydrate-active enzymes from marine bacteria to Japanese gut microbiota.

10 Crisp, A., *Genome Biol* (13 Mar 2015); 16 (1): 50. Expression of multiple horizontally acquired genes is a hallmark of both vertebrate and invertebrate genomes.

11 Cantarel, B.L., *PLoS One* (2012); 7(6): e28742. Complex carbohydrate utilization by the healthy human microbiome.

12 Georg, J.M., *Obes Rev* (Feb 2013); 14(2): 129–44. Review: efficacy of alginate supplementation in relation to appetite regulation and metabolic risk factors: evidence from animal and human studies.

13 Brown, E.S., *Nutr Rev* (Mar 2014); 72(3): 205–16. Seaweed and human health.

14 Hehemann, J.H., *Proc Natl Acad Sci* (27 Nov 2012); 109(48): 19786–91. Bacteria of the human gut microbiome catabolize red seaweed glycans with carbohydrate-active enzyme updates from extrinsic microbes.

15 Pirotta, M., *BMJ* (4 Sep 2004); 329(7465): 548. Effect of lactobacillus in preventing post-antibiotic vulvovaginal candidiasis: a randomised controlled trial.

16 Dey, B., *Curr HIV Res* (Oct 2013); 11(7): 576–94. Protein-based HIV-1 microbicides.

17 Wang, J., *J Nutr* (Jan 2014); 144(1): 98–105. Dietary supplementation with white button mushrooms augments the protective immune response to Salmonella vaccine in mice.

18 Varshney, J., *J Nutr* (Apr 2013); 143(4): 526–32. White button mushrooms increase microbial diversity and accelerate the resolution of *Citrobacter rodentium* infection in mice.

9 Protein: Milk Products

1 Campbell, T.C., *Am J Cardiol* (26 Nov 1998); 82(10B): 18T–21T. Diet, lifestyle, and the etiology of coronary artery disease: the Cornell China Study.

2 Minger, D., Dairy consumption in rural China; http://rawfoodsos. com/2010/07/07/the-china-study-fact-or-fallac/

3 Madani, S., *Nutrition* (May 2000); 16(5): 368–75. Dietary protein level and origin (casein and highly purified soybean protein) affect hepatic storage, plasma lipid transport, and antioxidative defense status in the rat.

4 Brüssow, H., *Environ Microbiol* (Aug 2013); 15(8): 2154–61. Nutrition, population growth and disease: a short history of lactose.

5 Savaiano, D.A., *Am J Clin Nutr* (May 2014); 99(5 Suppl): 1251S–5S. Lactose digestion from yogurt: mechanism and relevance.

6 Tishkoff, S.A., *Nat Genet* (2007); 39: 31–40. Convergent adaptation of human lactase persistence in Africa and Europe.

7 Spector, T., *Identically Different* (Weidenfeld & Nicolson, 2012)

8 Quigley, L., *FEMS Microbiol Rev* (Sep 2013); 37(5): 664–98. The complex microbiota of raw milk.

9 Bailey, R.K., *J Natl Med Assoc* (Summer 2013); 105(2): 112–27. Lactose intolerance and health disparities among African Americans and Hispanic Americans: an updated consensus statement.

10 Suchy, F.J., *NIH Consensus State Sci Statements* (24 Feb 2010); 27(2): 1–27. NIH consensus development conference statement: Lactose intolerance and health.

11 Petschow, B., *Ann NY Acad Sci* (Dec 2013); 1306: 1–17. Probiotics, prebiotics, and the host microbiome: the science of translation.

12 Prentice, A.M., *Am J Clin Nutr* (May 2014); 99 (5 Suppl): 1212S–16S. Dairy products in global public health.

13 Silventoinen, K., *Twin Res* (Oct 2003); 6(5): 399–408. Heritability of adult body height: a comparative study of twin cohorts in eight countries.

14 Wood, A.R., *Nat Genet* (5 Oct 2014); doi:10.1038/ng.3097. Defining the role of common variation in the genomic and biological architecture of adult human height.

15 Floud, R., *The Changing Body: New Approaches to Economic and Social History* (Cambridge University Press, 2011)

16 http://www.ers.usda.gov/media/1118789/err149.pdf

17 Quigley, L., *J Dairy Sci* (Aug 2013); 96(8): 4928–37. The microbial content of raw and pasteurized cow's milk as determined by molecular approaches.

18 Pottenger, F.M., *Pottenger's Cats: A Study in Nutrition* (Price Pottenger Nutrition, 1995)

19 Scher, J.U., *eLife* (Nov 2013); 5; 2: e01202. Expansion of intestinal *Prevotella copri* correlates with enhanced susceptibility to arthritis.

10 Carbohydrates: of which Sugars

1 http://www.telegraph.co.uk/news/worldnews/europe/netherlands/10314705/Sugar-is-addictive-and-the-most-dangerous-drug-of-the-times.html

2 Locke, A.E., *Nature* (12 Feb 2015); 518(7538): 197-206. Genetic studies of body mass index yield new insights for obesity biology.

3 Qi, Q., *N Engl J Med* (11 Oct 2012); 367(15): 1387–96. Sugar-sweetened beverages and genetic risk of obesity.

4 Keskitalo, K., *Am J Clin Nutr* (Aug 2008); 88(2): 263–71. The Three-Factor Eating Questionnaire, body mass index, and responses to sweet and salty fatty foods: a twin study of genetic and environmental associations.

5 Keskitalo, K., *Am J Clin Nutr* (Dec 2007); 86(6): 1663–9. Same genetic components underlie different measures of sweet taste preference.

6 http://www.who.int/nutrition/sugars_public_consultation/en/

7 https://www.gov.uk/government/uploads/system/uploads/attachment_data/file/470179/Sugar_reduction_the_evidence_for_action.pdf

8 Long, M.W., *Am J Prev Med* (Jul 2015); 49(1): 112–23. Cost effectiveness of a sugar-sweetened beverage excise tax in the US.

9 Yudkin, J., *Pure, White and Deadly*, reissue edn (Penguin, 2012)

10 Wilska, A., *Duodecim* (1947); 63: 449–510. Sugar caries – the most prevalent disease of our century.

11 Sheiham, A., *Int J Epidemiol* (Jun 1984); 13(2): 142–7. Changing trends in dental caries.

12 Birkeland, J.M., *Caries Res* (Mar–Apr 2000); 34(2): 109–16. Some factors associated with the caries decline among Norwegian children and adolescents: age-specific and cohort analyses.

13 Masadeh, M., *J Clin Med Res* (2013); 5.5: 389–94. Antimicrobial activity of common mouthwash solutions on multidrug-resistance bacterial biofilms.

14   Kapil, V., *Free Radic Biol Med* (Feb 2013); 55: 93–100. Physiological role for nitrate-reducing oral bacteria in blood pressure control.

15   Fine, D.H., *Infect Immun* (May 2013); 81(5): 1596–605. A lactotransferrin single nucleotide polymorphism demonstrates biological activity that can reduce susceptibility to caries.

16   Holz, C., *Probiotics Antimicrob Proteins* (2013); 5: 259–63. *Lactobacillus paracasei* DSMZ16671 reduces mutans streptococci: a short-term pilot study.

17   Teanpaisan, R., *Clin Oral Investig* (Apr 2014); 18(3): 857–62. *Lactobacillus paracasei* SD1, a novel probiotic, reduces mutans streptococci in human volunteers: a randomized placebo-controlled trial.

18   Glavina, D., *Coll Antropol* (Mar 2012); 36(1): 129–32. Effect of LGG yoghurt on *Streptococcus mutans* and Lactobacillus spp. salivary counts in children.

19   Marcenes, W., *J Dent Res* (Jul 2013); 92(7): 592–7. Global burden of oral conditions in 1990–2010: a systematic analysis.

20   Bernabé, E., *Am J Public Health* (Jul 2014); 104(7): e115–21. Age, period and cohort trends in caries of permanent teeth in four developed countries.

21   Bray, G.A., *Am J Clin Nutr* (2004); 79: 537–43. Consumption of high-fructose corn syrup in beverages may play a role in the epidemic of obesity.

22   Ng, S.W., *Br J Nutr* (Aug 2012); 108(3): 536–51. Patterns and trends of beverage consumption among children and adults in Great Britain, 1986–2009.

23   Bray, G.A., *Am J Clin Nutr* (2004); 79: 537–43. Consumption of high-fructose corn syrup in beverages may play a role in the epidemic of obesity.

24   Hu, F.B., *Physiol Behav* (2010); 100: 47–54. Sugar-sweetened beverages and risk of obesity and Type 2 diabetes: epidemiologic evidence.

25   Mitsui, T., *J Sports Med Phys Fitness* (Mar 2001); 41(1): 121–3. Colonic fermentation after ingestion of fructose-containing sports drink.

26   Bergheim, I., *J Hepatol* (Jun 2008); 48(6): 983–92. Antibiotics protect against fructose-induced hepatic lipid accumulation in mice: role of endotoxin.

27   Bray, G.A., *Diabetes Care* (Apr 2014); 37(4): 950–6. Dietary sugar and body weight: have we reached a crisis in the epidemic of obesity and diabetes?: health be damned! Pour on the sugar.

28   de Ruyter, J.C., *N Engl J Med* (11 Oct 2012); 367(15): 1397–406. A trial of sugar-free or sugar-sweetened beverages and body weight in children.

29   Sartorelli, D.S., *Nutr Metab Cardiovasc Dis* (Feb 2009); 19(2): 77–83. Dietary fructose, fruits, fruit juices and glucose tolerance status in Japanese–Brazilians.

30 van Buul, V.J., *Nutr Res Rev* (Jun 2014); 27(1): 119–30. Misconceptions about fructose-containing sugars and their role in the obesity epidemic.

31 Kahn, R., *Diabetes Care* (Apr 2014); 37(4): 957–62. Dietary sugar and body weight: have we reached a crisis in the epidemic of obesity and diabetes?: we have, but the pox on sugar is overwrought and overworked.

32 Te Morenga, L., *BMJ* (15 Jan 2012); 346: e7492. Dietary sugars and body weight: systematic review and meta-analyses of randomised controlled trials and cohort studies.

33 Sievenpiper, J.L., *Ann Intern Med* (21 Feb 2012); 156: 291–304. Effect of fructose on body weight in controlled feeding trials: a systematic review and meta-analysis.

34 Kelishadi, R., *Nutrition* (May 2014); 30: 503–10. Association of fructose consumption and components of metabolic syndrome in human studies: a systematic review and meta-analysis.

11 Carbohydrates: Non-sugars

1 Claesson, M.J., *Nature* (9 Aug 2012); 488(7410): 178–84. Gut microbiota composition correlates with diet and health in the elderly.

2 van Tongeren, S., *Appl. Environ. Microbiol* (2005); 71(10): 6438–42. Fecal microbiota composition and frailty.

3 Friedman, M., *J Agric Food Chem* (9 Oct 2013); 61(40): 9534–50. Anticarcinogenic, cardioprotective, and other health benefits of tomato compounds lycopene, α-tomatine, and tomatidine in pure form and in fresh and processed tomatoes.

4 http://www.dailymail.co.uk/femail/article-2692758/Diet-guru-FreeLee-Banana-Girl-fire-controversial-views-claims-chemo-kills-losing-period-good-you.html

5 Mosley, M., and Spencer, M., *The Fast Diet* (Short Books, 2013)

6 Bao, Q., *Mol Cell Endocrinol*, 16 Jul 2014; 394(1–2):115–18. Ageing and age-related diseases – from endocrine therapy to target therapy.

7 Hankey, C., *FASEB J* (Apr 2015); 29(1): 117.4. A systematic review of the literature on intermittent fasting for weight management.

8 Johnson, J.B., *Med Hypotheses*, 2006; 67:209–11. The effect on health of alternate day calorie restriction: eating less and more than needed on alternate days prolongs life. Vallejo, E.A., *Rev Clin Esp*, 63 (1956): 25–7. La dieta de hambre a días alternos en la alimentación de los viejos.

9 Nicholson, A., *Soc Sci Med* (2009); 69 (4): 519–28. Association between attendance at religious services and self-reported health in 22 European countries. Eslami, S., *Bioimpacts* (2012); 2(4): 213–15. Annual fasting; the early calories restriction for cancer prevention.

10 Carey, H.V., *Am J Physiol Regul Integr Comp Physiol* (1 Jan 2013); 304(1): R33–42 Seasonal restructuring of the ground squirrel gut microbiota over the annual hibernation cycle.

11   Costello, E.K., *ISME J* (Nov 2010); 4(11):1375–85. Postprandial remodelling of the gut microbiota in Burmese pythons.

12   Zarrinpar, A., *Cell Metab* (2 Dec 2014); 20(6):1006–17. doi:10.1016/j. cmet.2014.11.008. Diet and feeding pattern affect the diurnal dynamics of the gut microbiome.

13   Remely, M., *Wien klin Wochenschr* (Mar 2015) 127: 394–8. Increased gut microbiota diversity and abundance of *Faecalibacterium prausnitzii* and *Akkermansia* after fasting: a pilot study.

14   Casazza, K., *N Engl J Med* (31 Jan 2013); 368(5): 446–54. Myths, presumptions, and facts about obesity.

15   Betts, J.A., *Am J Clin Nutr* (4 Jun 2014); 100(2): 539–47. The causal role of breakfast in energy balance and health: a randomized controlled trial in lean adults. Dhurandhar, E.J., *Am J Clin Nutr*, (4 Jun 2014); 100(2): 507–13. The effectiveness of breakfast recommendations on weight loss: a randomized controlled trial.

16   de la Hunty, A., *Obes Facts* (2013); 6(1): 70–85. Does regular breakfast cereal consumption help children and adolescents stay slimmer? A systematic review and meta-analysis.

17   Brown, A.W., *Am J Clin Nutr* (Nov 2013); 98(5): 1298–308. Belief beyond the evidence: using the proposed effect of breakfast on obesity to show two practices that distort scientific evidence.

18   Desai, A.V., *Twin Res* (Dec 2004); 7(6): 589–95. Genetic influences in self-reported symptoms of obstructive sleep apnoea and restless legs: a twin study.

19   Shelton, H., *Hygienic systems*, vol. II, *Health Research*. (Pomeroy, WA, 1934).

20   Stote, K.S., *Am J Clin Nutr* (Apr 2007); 85(4): 981–8. A controlled trial of reduced meal frequency without caloric restriction in healthy, normal-weight, middle-aged adults.

21   Di Rienzi, S.C., *eLife* (1 Oct 2013); 2: e01102.doi:10.7554/eLife.01102. The human gut and groundwater harbor non-photosynthetic bacteria belonging to a new candidate phylum sibling to Cyanobacteria.

22   Watanabe, F., *Nutrients* (5 May 2014); 6(5): 1861–73. Vitamin B12-containing plant food sources for vegetarians.

23   Brown, M.J., *Am J Clin Nutr* (2004); 80: 396–403. Carotenoid bioavailability is higher from salads ingested with full-fat than with fat-reduced salad dressings as measured with electrochemical detection.

24   Sonnenburg, E.D., *Cell Metab* (20 Aug 2014); pii: S1550-4131(14)00311-8. Starving our microbial self: the deleterious consequences of a diet deficient in microbiota-accessible carbohydrates.

25   Wertz, A.E., *Cognition* (Jan 2014); 130(1): 44–9. Thyme to touch: infants possess strategies that protect them from dangers posed by plants.

26 Knaapila, A., *Physiol Behav* (15 Aug 2007); 91(5): 573–8. Food neophobia shows heritable variation in humans.

12 Fibre

1 Anderson, J.C., *Am J Gastroenterol* (Oct 2014); 109(10): 1650–2. Editorial: constipation and colorectal cancer risk: a continuing conundrum.

2 Threapleton, D.E., *BMJ* (19 Dec 2013); 347: f6879. Dietary fibre intake and risk of cardiovascular disease: systematic review and meta-analysis.

3 Kim, Y., *Am J Epidemiol* (15 Sep 2014);180(6): 565–73. Dietary fiber intake and total mortality: a meta-analysis of prospective cohort studies.

4 Thies, F., *Br J Nutr* (Oct 2014); 112 Suppl 2: S19–30. Oats and CVD risk markers: a systematic literature review.

5 Musilova, S., *Benef Microbes* (Sep 2014); 5(3): 273–83. Beneficial effects of human milk oligosaccharides on gut microbiota.

6 Ukhanova, M., *Br J Nutr* (28 Jun 2014); 111(12): 2146–52. Effects of almond and pistachio consumption on gut microbiota composition in a randomised cross-over human feeding study.

7 Dunn, S., *Eur J Clin Nutr* (Mar 2011); 65(3): 402–8. Validation of a food frequency questionnaire to measure intakes of inulin and oligofructose.

8 Moshfegh, A.J., *J Nutr* (1999); 129: 1407S–11S. Presence of inulin and oligofructose in the diets of Americans.

9 van Loo, J., *Crit Rev Food Sci Nutr* (Nov 1995); 35(6): 525–52. On the presence of inulin and oligofructose as natural ingredients in the Western diet.

10 Teucher, B., *Twin Res Hum Genet* (Oct 2007); 10(5): 734–48. Dietary patterns and heritability of food choice in a UK female twin cohort.

11 Williams, F.M., *BMC Musculoskelet Disord* (8 Dec 2010); 11: 280. Dietary garlic and hip osteoarthritis: evidence of a protective effect and putative mechanism of action.

12 Lissiman, E., *Cochrane Database Syst Rev* (11 Nov 2014);11: CD006206. Garlic for the common cold; Josling P., *Advances in therapy* (2001);18(4):189–93. Preventing the common cold with a garlic supplement: a double-blind, placebo-controlled survey.

13 Zeng, T., *J Sci Food Agric* (2012); 92 (9): 1892–1902. A meta-analysis of randomized, double-blind, placebo-controlled trials for the effects of garlic on serum lipid profile.

14 O'Brien, C.L., *PLoS One* (1 May 2013); 8(5): e62815. Impact of colonoscopy bowel preparation on intestinal microbiota.

15 Kellow, N.J., *Br J Nutr* (14 Apr 2014); 111(7): 1147–61. Metabolic benefits of dietary prebiotics in human subjects: a systematic review of randomised controlled trials.

16 Dewulf, E.M., *Gut* (Aug 2013); 62(8): 1112–21. Insight into the prebiotic concept: lessons from an exploratory, double-blind intervention study

with inulin-type fructans in obese women.

17 Salazar, N., *Clin Nutr* (11 Jun 2014); pii: S0261-5614(14)00159-9.doi: 10.1016/j.clnu.2014.06.001. Inulin-type fructans modulate intestinal Bifidobacteria species populations and decrease fecal short-chain fatty acids in obese women.

18 http://www.jigsawhealth.com/supplements/butyrex

19 Davis, W., *Wheat Belly* (Rodale, 2011)

20 Murray, J., *Am J Clin Nutr* (2004); 79(4) 669–73. Effect of a gluten-free diet on gastrointestinal symptoms in celiac disease.

21 Perry, G.H., *Nature Genetics* (2007); 39: 1256–60. Diet and the evolution of human amylase gene copy number variation.

22 Falchi, M., *Nature Genetics* (May 2014); 46(5): 492–7. Low copy number of the salivary amylase gene predisposes to obesity.

23 Claussnitzer, M., *N Engl J Med* (3 Sep 2015); 373(10): 895–907. FTO obesity variant circuitry and adipocyte browning in humans.

24 Staudacher, H.M., *Nat Rev Gastroenterol Hepatol* (Apr 2014);11(4): 256–66. Mechanisms and efficacy of dietary FODMAP restriction in IBS.

13 Artificial Sweeteners and Preservatives

1 Di Salle, F., *Gastroenterology* (Sep 2013); 145(3): 537–9. Effect of carbonation on brain processing of sweet stimuli in humans.

2 de Ruyter, J.C., *N Engl J Med* (11 Oct 2012); 367(15): 1397–406. A trial of sugar-free or sugar-sweetened beverages and body weight in children.

3 de Ruyter, J.C., *PLoS One* (22 Oct 2013); 8(10): e78039. The effect of sugar-free versus sugar-sweetened beverages on satiety, liking and wanting: an 18-month randomized double-blind trial in children.

4 Nettleton, J.A., *Diabetes Care* (2009); 32(4): 688–94. Diet soda intake and risk of incident metabolic syndrome and Type 2 diabetes in the Multi-Ethnic Study of Atherosclerosis (MESA).

5 Lutsey, P.L., *Circulation* (12 Feb 2008); 117(6): 754–61. Dietary intake and the development of the metabolic syndrome: the Atherosclerosis Risk in Communities study.

6 Hill, S.E., *Appetite* (13 Aug 2014); pii: S0195-6663(14)00400-0. The effect of non-caloric sweeteners on cognition, choice, and post-consumption satisfaction.

7 Bornet, F.J., *Appetite* (2007); 49(3): 535–53. Glycaemic response to foods. Impact on satiety and long-term weight regulation.

8 Schiffman, S.S., *J Toxicol Environ Health B Crit Rev* (2013); 16(7): 399–451. Sucralose, a synthetic organochlorine sweetener: overview of biological issues.

9 Green, E., *Physiol Behav* (5 Nov 2012); 107(4): 560–7. Altered processing of sweet taste in the brain of diet soda drinkers.

10 Pepino, M.Y., *Diabetes Care* (2013); 36: 2530–5. Sucralose affects glycemic and hormonal responses to an oral glucose load.

11 Abou-Donia, M.B., *J Toxicol Environ Health A* (2008); 71(21): 1415–29. Splenda alters gut microflora and increases intestinal p-glycoprotein and cytochrome p-450 in male rats.

12 Wu, G.D., *Science* (7 Oct 2011); 334(6052): 105–8. Linking long-term dietary patterns with gut microbial enterotypes.

13 Jackson, M., *Gut* (2016) http://gut.bmj.com/content/early/2015/12/30/gutjnl-2015-3108bl.abstract. Proton pump inhibitors after the composition of the gut microbiota.

14 Gostner, J., *Curr Pharm Des* (2014); 20(6): 840–9. Immunoregulatory impact of food antioxidants.

14 Contains Cocoa and Caffeine

1 Shin, S.Y., *Nat Genet* (Jun 2014); 46(6): 543–50. An atlas of genetic influences on human blood metabolites.

2 Mayer, E.A., *J Neurosci* (2014), 34(46): 15490–6. Gut microbes and the brain: paradigm shift in neuroscience.

3 Ellam, S., *Ann Rev Nutr* (2013); 33: 105–28. Cocoa and human health.

4 Stafford, L.D., *Chem Senses* (Mar 13 2015); bj007, p.ii. Obese individuals have higher preference and sensitivity to odor of chocolates.

5 Ellam, S., *Ann Rev Nutr* (2013); 33: 105–28. Cocoa and human health

6 Golomb, B.A., *Arch Intern Med* (26 Mar 2012); 172(6): 519–21. Association between more frequent chocolate consumption and lower body mass index.

7 Khan, N., *Nutrients* (21 Feb 2014); 6(2): 844–80.doi:10.3390/nu6020844. Cocoa polyphenols and inflammatory markers of cardiovascular disease.

8 Jennings, A., *J Nutr* (Feb 2014); 144(2): 202–8. Intakes of anthocyanins and flavones are associated with biomarkers of insulin resistance and inflammation in women.

9 Tzounis, X., *Am J Clin Nutr* (Jan 2011); 93(1): 62–72. Prebiotic evaluation of cocoa-derived flavanols in healthy humans by using a randomized, controlled, double-blind, crossover intervention study.

10 http://www.cocoavia.com/how-do-i-use-it/ingredients-nutritional-information

11 Martin, F.P., *J Proteome Res* (7 Dec 2012); 11(12): 6252–63. Specific dietary preferences are linked to differing gut microbial metabolic activity in response to dark chocolate intake.

12 Esser, D., *FASEB J* (Mar 2014); 28(3): 1464–73. Dark chocolate consumption improves leukocyte adhesion factors and vascular function in overweight men.

13 Moco, S., *J Proteome Res* (5 Oct 2012); 11(10): 4781–90. Metabolomics view on gut microbiome modulation by polyphenol-rich foods.

14 Langer, S., *J Agric Food Chem* (10 Aug 2011); 59(15): 8435–41. Flavanols and methylxanthines in commercially available dark chocolate: a study of the correlation with non-fat cocoa solids.

15 Zhang, C., *Eur J Epidemiol* (30 Oct 2014); Tea consumption and risk of cardiovascular outcomes and total mortality: a systematic review and meta-analysis of prospective observational studies.

16 Ludwig, I.A., *Food Funct* (Aug 2014); 5(8): 1718–26. Variations in caffeine and chlorogenic acid contents of coffees: what are we drinking?

17 Crippa, A., *Am J Epidemiol* (24 Aug 2014); pii: kwu194. Coffee consumption and mortality from all causes, cardiovascular disease, and cancer: a dose-response meta-analysis.

18 Denoeud, F., *Science* (2014); 345: 1181–4. The coffee genome provides insight into the convergent evolution of caffeine biosynthesis.

19 http://www.dailymail.co.uk/health/article-3027/How-healthy-cup-coffee.html

20 Teucher, B., *Twin Res Hum Genet* (Oct 2007); 10(5): 734–48. Dietary patterns and heritability of food choice in a UK female twin cohort.

21 Amin, N., *Mol Psychiatry* (Nov 2012); 17(11): 1116–29. Genome-wide association analysis of coffee drinking suggests association with CYP1A1/CYP1A2 and NRCAM.

22 Coelho, C., *J Agric Food Chem* (6 Aug 2014); 62(31): 7843–53. Nature of phenolic compounds in coffee melanoidins.

23 Gniechwitz, D., *J Agric Food Chem* (22 Aug 2007); 55(17): 6989–96. Dietary fiber from coffee beverage: degradation by human fecal microbiota.

24 Vinson, J.A., *Diabetes Metab Syndr Obes* (2012); 5: 21–7. Randomized, double-blind, placebo-controlled, linear dose, crossover study to evaluate the efficacy and safety of a green coffee bean extract in overweight subjects.

15 Contains Alcohol

1 Rehm, J., *Lancet* (26 Apr 2014); 383(9927): 1440–2. Russia: lessons for alcohol epidemiology and alcohol policy.

2 Spector, T., *Identically Different* (Weidenfeld & Nicolson, 2012)

3 Peng, Y., *BMC Evol Biol* (20 Jan 2010); 10: 15. The ADH1B Arg47His polymorphism in east Asian populations and expansion of rice domestication in history.

4 Criqui, M.H., *Lancet* (1994); 344; 1719-23. Does diet or alcohol explain the French paradox?

5 Di Castelnuovo, A., *Arch Intern Med* (11–25 Dec 2006); 166(22): 2437–45. Alcohol dosing and total mortality in men and women: an updated meta-analysis of 34 prospective studies.

6 Gepner Y., and R. Golan, *Ann Intern Med* (20 Oct 2015); 163(8): 569–79. Effects of initiating moderate alcohol intake on cardiometric risk in adults with Type 2 diabetes.

7 Chawla, R., *BMJ* (4 Dec 2004); 329 (7478): 1308. Regular drinking might explain the French paradox.

8 Spitaels, F., *PLoS One* (18 Apr 2014); 9(4): e95384. The microbial diversity of traditional spontaneously fermented lambic beer.

9 Vang, O., *Ann NY Acad Sci* (Jul 2013); 1290: 1–11. What is new for resveratrol? Is a new set of recommendations necessary?

10 Tang, P.C., *Pharmacol Res* (22 Aug 2014); pii: S1043-6618(14)00138-8. Resveratrol and cardiovascular health – Promising therapeutic or hopeless illusion?

11 Wu, G.D., *Science* (7 Oct 2011); 334(6052): 105–8. Linking long-term dietary patterns with gut microbial enterotypes.

12 Queipo-Ortuño, M.I., *Am J Clin Nutr* (Jun 2012); 95(6): 1323–34. Influence of red wine polyphenols and ethanol on the gut microbiota ecology and biochemical biomarkers.

13 Chiva-Blanch, G., *Alcohol* (May–Jun 2013); 48(3): 270–7. Effects of wine, alcohol and polyphenols on cardiovascular disease risk factors: evidence from human studies.

14 Bala, S., *PLoS One* (14 May 2014); 9(5). Acute binge drinking increases serum endotoxin and bacterial DNA levels in healthy individuals.

15 Blednov, Y.A., *Brain Behav Immun* (Jun 2011); 25 Suppl 1: S92–S105. Activation of inflammatory signaling by lipopolysaccharide produces a prolonged increase of voluntary alcohol intake in mice.

16 Knott, C.S., *BMJ* (2015); 350: h384. All-cause mortality and the case for age-specific consumption guidelines.

17 Ferrari, P., *Eur J Clin Nutr* (Dec 2012); 66(12): 1303–8. Alcohol dehydrogenase and aldehyde dehydrogenase gene polymorphisms, alcohol intake and the risk of colorectal cancer in the EPIC study.

18 Holmes, M.V., *BMJ* (2014); 349: g4164. Association between alcohol and cardiovascular disease: Mendelian randomisation analysis based on individual participant data.

16 Vitamins

1 Goto, Y., *Immunity* (17 Apr 2014); 40(4): 594–607. Segmented filamentous bacteria antigens presented by intestinal dendritic cells drive mucosal Th17 cell differentiation.

2 Rimm, E.B., *N Engl J Med* (20 May 1993); 328(20): 1450–6. Vitamin E consumption and the risk of coronary heart disease in men.

3 Lippman, S.M., *JAMA* (2009); 301(1): 39–51. Effect of selenium and vitamin E on risk of prostate cancer and other cancers.

4 Guallar, E., *Ann Intern Med* (17 Dec 2013); 159(12): 850–1. Enough is

enough: Stop wasting money on vitamin and mineral supplements.

5 Bjelakovic, G., *Cochrane Database Syst Rev* (14 Mar 2012); 3: CD007176. Antioxidant supplements for prevention of mortality in healthy participants and patients with various diseases.

6 Risk and Prevention Study Collaborative Group, *N Engl J Med* (9 May 2013); 368(19): 1800–8.n-3. Fatty acids in patients with multiple cardiovascular risk factors.

7 Qin, X., *Int J Cancer* (1 Sep 2013); 133(5): 1033–41. Folic acid supplementation and cancer risk: a meta-analysis of randomized controlled trials.

8 Burdge, G.C., *Br J Nutr* (14 Dec 2012); 108(11): 1924–30. Folic acid supplementation in pregnancy: Are there devils in the detail?

9 Qin, X., *Clin Nutr* (Aug 2014); 33(4): 603–12. Folic acid supplementation with and without vitamin B6 and revascularization risk: a meta-analysis of randomized controlled trials.

10 Murto, T., *Acta Obstet Gynecol Scand* (Jan 2015); 94(1): 65–71. Folic acid supplementation and methylenetetrahydrofolate reductase (MTHFR) gene variations in relation to in vitro fertilization pregnancy outcome.

11 Huang, Y., *Int J Mol Sci* (14 Apr 2014); 15(4): 6298–313. Maternal high folic acid supplement promotes glucose intolerance and insulin resistance in male mouse offspring fed a high-fat diet.

12 Clarke, J.D., *Pharmacol Res* (Nov 2011); 64(5): 456-63.Bioavailability and inter-conversion of sulforaphane and erucin in human subjects consuming broccoli sprouts or broccoli supplement in a cross-over study design.

13 Bolland, M.J., *J Bone Miner Res* (11 Sep 2014); doi:10.1002/jbmr.2357. Calcium supplements increase risk of myocardial infarction.

14 Moyer, V.A., *Ann Intern Med* (2013); 158(9): 691–6. Vitamin D and calcium supplementation to prevent fractures in adults: U.S. preventive services task force recommendation statement.

15 Chang, Y.M., *Int J Epidemiol* (Jun 2009); 38(3): 814–30. Sun exposure and melanoma risk at different latitudes: a pooled analysis of 5700 cases and 7216 controls.

16 Bataille, V., *Med Hypotheses* (Nov 2013); 81(5): 846–50. Melanoma. Shall we move away from the sun and focus more on embryogenesis, body weight and longevity?

17 Gandini, S., *PLoS One* (2013); 8(11): e78820. Sunny holidays before and after melanoma diagnosis are respectively associated with lower Breslow thickness and lower relapse rates in Italy.

18 Hunter, D., *J Bone Miner Res* (Nov 2000); 15(11): 2276–83. A randomized controlled trial of vitamin D supplementation on preventing postmenopausal bone loss and modifying bone metabolism using identical twin pairs.

19 Schöttker, B., *BMJ* (17 Jun 2014); 348: g3656.doi:10.1136/bmj.g3656.

Vitamin D and mortality: meta-analysis of individual participant data from a large consortium of cohort studies from Europe and the United States.

20 Sanders, K.M., *Calcif Tissue Int* (Feb 2013); 92(2): 191–206. Is high dose vitamin D harmful?

21 Bischoff-Ferrari, H.A., *JAMA* (Jan 2016) http://archinte.jamanetwork.com/article.aspx?articleid=2478897. Monthly high-dose vitamin D treatment for the prevention of functional decline.

22 Bjelakovic, G., *Cochrane Database Syst Rev* (23 Jun 2014); 6: CD007469. doi: 10.1002/14651858.CD007469.pub2. Vitamin D supplementation for prevention of cancer in adults.

23 Afzal, S., *BMJ* (18 Nov 2014); 349: g6330.doi:10.1136/bmj.g6330. Genetically low vitamin D concentrations and increased mortality: Mendelian randomisation analysis in three large cohorts.

24 Age-related Eye Disease Study 2 Research Group, *JAMA* (15 May 2013); 309(19): 2005–15. Lutein + zeaxanthin and omega-3 fatty acids for age-related macular degeneration: the Age-related Eye Disease Study 2 (AREDS2) randomised clinical trial.

25 Degnan, P.H., *Cell Metab* (4 Nov 2014); 20(5): 769–74. Vitamin B12 as a modulator of gut microbial ecology.

17 Warning: May Contain Antibiotics

1 Shapiro, D.J., *J Antimicrob Chemother* (25 Jul 2013); Antibiotic prescribing for adults in ambulatory care in the USA, 2007–09.

2 Hicks, L., *N Engl J Med* (2013); 368; 1461–2. U.S. outpatient antibiotic prescribing, 2010

3 Garrido, D., *Microbiology* (Apr 2013); 159(Pt 4): 649–64. Consumption of human milk glycoconjugates by infant-associated bifidobacteria: mechanisms and implications.

4 Baaqeel, H., *BJOG* (May 2013); 120(6): 661–9.doi:10.1111/1471-0528.12036. Timing of administration of prophylactic antibiotics for caesarean section: a systematic review and meta-analysis.

5 Kozhimannil, K.B., *PLoS Med* (21 Oct 2014); 11(10): e1001745. Maternal clinical diagnoses and hospital variation in the risk of Cesarean delivery: analyses of a national US hospital discharge database.

6 Kozhimannil, K.B., *Health Aff* (Millwood) (Mar 2013); 32(3): 527–35. Cesarean delivery rates vary tenfold among US hospitals; reducing variation may address quality and cost issues.

7 Zhang, J., *Obstet Gynecol* (2008); 111: 1077–82. Cesarean delivery on maternal request in Southeast China.

8 Dominguez-Bello, M., *Proc Natl Acad Sci USA* (29 Jun 2010); 107(26): 11971–5. Delivery mode shapes the acquisition and structure of the initial microbiota across multiple body habitats in newborns.

9 Grönlund, M.M., *J Pediatr Gastroenterol Nutr* (1999); 28: 19–25. Fecal microflora in healthy infants born by different methods of delivery: Permanent changes in intestinal flora after Cesarean delivery.

10 Cho, C.E., *Am J Obstet Gynecol* (Apr 2013); 208(4): 249–54. Cesarean section and development of the immune system in the offspring.

11 Thavagnanam, S., *Clin Exp Allergy* (Apr 2008); 38(4): 629–33. A meta-analysis of the association between caesarean section and childhood asthma.

12 Gough, E.K., *BMJ* (15 Apr 2014); 348: g2267. The impact of antibiotics on growth in children in low and middle income countries: systematic review and meta-analysis of randomised controlled trials.

13 Blaser, M., *Missing Microbes* (Henry Holt, 2014)

14 http://www.cdc.gov/obesity/data/adult.html

15 Trasande, L., *Int J Obes* (Jan 2013); 37(1): 16–23. Infant antibiotic exposures and early-life body mass.

16 Ajslev, T.A., *Int J Obes* 2011; 35: 522–9. Childhood overweight after establishment of the gut microbiota: the role of delivery mode, pre-pregnancy weight and early administration of antibiotics.

17 Bailey, L.C., *JAMA Pediatr* (29 Sep 2014); doi:10.1001/jamapediatrics. Association of antibiotics in infancy with early childhood obesity.

18 Nobel, Y.R., *Nature Comms* (30 Jun 2015); 6: 7486. Metabolic and metagenomic outcomes from early-life pulsed antibiotic treatment.

19 Darmasseelane, K., *PLoS One* (2014); 9(2): e87896.doi:10.1371. Mode of delivery and offspring body mass index, overweight and obesity in adult life: a systematic review and meta-analysis.

20 http://www.wired.com/wiredscience/2010/12/news-update-farm-animals-get-80-of-antibiotics-sold-in-us/

21 Visek, W.J., *J Animal Sciences* (1978); 46; 1447–69.The mode of growth promotion by antibiotics.

22 Pollan, M., *The Omnivore's Dilemma* (Bloomsbury, 2007)

23 http://www.dutchnews.nl/news/archives/2014/06/illegal_antibiotics_found_on_f/

24 Burridge, L., *Aquaculture* (2010); Elsevier BV 306 (1–4): 7–23 Chemical use in salmon aquaculture: a review of current practices and possible environmental effects.

25 Karthikeyan, K.G., *Sci Total Environ* (15 May 2006); 361(1–3). Occurrence of antibiotics in wastewater treatment facilities in Wisconsin, USA.

26 Jiang, L., *Sci Total Environ* (1 Aug 2013); 458–460: 267–72.doi. Prevalence of antibiotic resistance genes and their relationship with antibiotics in the Huangpu River and the drinking water sources, Shanghai, China.

27 Huerta, B., *Sci Total Environ* (1 Jul 2013); 456–7: 161–70. Exploring the links between antibiotic occurrence, antibiotic resistance, and bacterial communities in water supply reservoirs.

28 Falcone-Dias, M.F., *Water Res* (Jul 2012); 46(11): 3612–22. Bottled mineral water as a potential source of antibiotic-resistant bacteria.

29 Cox, L.M., *Cell* (2014); 158: 705–21. Altering the intestinal microbiota during a critical development window has lasting metabolic consequences.

30 Gendrin, M., *Nature Communications* (6 Jan 2015); 6: 592. Antibiotics in ingested human blood affect the mosquito microbiota and capacity to transmit malaria.

31 Goldenberg, J.Z., *Cochrane Database Syst Rev* (31 May 2013); 5: CD006095. Probiotics for the prevention of Clostridium difficile-associated diarrhea in adults and children.

32 http://www.britishgut.org and http://www.americangut.org

18 Warning: May Contain Nuts

1 http://www.dailymail.co.uk/news/article-2724684/Nut-allergy-girl-went-anaphylactic-shock-plane-passenger-ignored-three-warnings-not-eat-nuts-board.html

2 Vinson, J.A., *Food Funct* (Feb 2012); 3(2): 134–40. Nuts, especially walnuts, have both anti-oxidant quantity and efficacy and exhibit significant potential health benefits.

3 Bes-Rastrollo, M., *Am J Clin Nutr* (2009); 89: 1913–19. Prospective study of nut consumption, long-term weight change, and obesity risk in women.

4 Estruch, R., *N Engl J Med* (4 Apr 2013); 368(14): 1279–90.

5 van den Brandt, P.A., *Int J Epidemiol* (Jun 2015); 44(3): 1038–49. Relationship of tree nut, peanut and peanut butter intake with total and cause-specific mortality: a cohort study and meta-analysis.

6 Schloss, O., *Arch Paed* (1912); 29: 219. A case of food allergy.

7 Golbert, T.M., *J Allergy* (Aug 1969); 44(2): 96–107. Systemic allergic reactions to ingested antigens.

8 West, C.E., *Curr Opin Clin Nutr Metab Care* (May 2014); 17(3): 261–6. Gut microbiota and allergic disease: new findings.

9 Du Toit, G., *J Allergy Clin Immunol* (Nov 2008); 122(5): 984–91. Early consumption of peanuts in infancy is associated with a low prevalence of peanut allergy.

10 Storrø, O., *Curr Opin Allergy Clin Immunol* (Jun 2013); 13(3): 257–62. Diversity of intestinal microbiota in infancy and the risk of allergic disease in childhood.

11 Ismail, I.H., *Pediatr Allergy Immunol* (Nov 2012); 23(7): 674–81. Reduced gut microbial diversity in early life is associated with later development of eczema.

12 Marrs, T., *Pediatr Allergy Immunol* (Jun 2013); 24(4): 311–20.e8. Is there an association between microbial exposure and food allergy? A systematic review.

13 Hesselmar, B., *Pediatrics* (Jun 2013); 131(6): e1829–37. Pacifier cleaning
    practices and risk of allergy development.

14 Strachan, D.P., *BMJ* (1989); 299: 1259–60. Hay fever, hygiene, and
    household size.

15 Holbreich, M., *J Allergy Clin Immunol* (Jun 2012); 129(6): 1671–3. Amish
    children living in northern Indiana have a very low prevalence of allergic
    sensitization.

16 Zupancic, M.L., *PLoS One* (2012); 7(8): e43052. Analysis of the gut
    microbiota in the Old Order Amish and its relation to the metabolic
    syndrome.

17 Roederer, M., *Cell* (2015); The genetic architecture of the human immune
    system.

18 Schaub, B., *J Allergy Clin Immunol* (Jun 2008); 121(6): 1491–9, 1499.e-13.
    Impairment of T-regulatory cells in cord blood of atopic mothers.

19 Smith, P.M., *Science* (2013); 341; 6145: 569–73. The microbial
    metabolites, short-chain fatty acids, regulate colonic Treg cell
    homeostasis.

20 Hansen, C.H., *Gut Microbes* (May–Jun 2013); 4(3): 241–5. Customizing
    laboratory mice by modifying gut microbiota and host immunity in an
    early 'window of opportunity'.

21 Wells, A.D., *Int Immunopharmacol* (31 Jul 2014); pii: S1567–9. Influence
    of farming exposure on the development of asthma and asthma-like
    symptoms.

22 Heinritz, S.N., *Nutr Res Rev* (Dec 2013); 26(2): 191–209. Use of pigs as
    a potential model for research into dietary modulation of the human gut
    microbiota.

23 Song, S.J., *eLife* (16 Apr 2013); 2:e00458. Cohabiting family members
    share microbiota with one another and with their dogs.

19 Best-before Date

1   Gormley, F., *J Epidemiol Infect* (May 2011); 139(5): 688–99. A 17-year
    review of food-borne outbreaks: describing the continuing decline in
    England and Wales (1992–2008).

2   Lyon, R.C., *J Pharm Sci* (2006); 95: 1549–60. Stability profiles of drug
    products extended beyond labelled expiration dates.

3   Khan, S.R., *J Pharm Sci* (May 2014); 103(5): 1331–6. United States Food
    and Drug Administration and Department of Defense shelf-life extension
    program of pharmaceutical products: progress and promise.

4   Mesnage, R., *Food Chem Toxicol* (Oct 2015); 84: 133–53. Potential toxic
    effects of glyphosate and its commercial formulations below regulatory
    limits.

5   http://www.iuss.org/19th%20WCSS/Symposium/pdf/1807.pdf

6   Lu, C., *Environ Health Perspect* (2006); 114: 260–3. Organic diets

significantly lower children's dietary exposure to organophosphorus pesticides.

Conclusion: The Checkout

1 http://www.britishgut.org and http://www.americangut.org
2 Youngster, I., *JAMA* (5 Nov 2014); 312(17): 1772–8. Oral, capsulized, frozen, fecal microbiota transplantation for relapsing Clostridium difficile infection.
3 http://www.openbiome.org/practitioner-map/
4 Spector, T.D., and R. Knight, *BMJ* (2015); 351: h5149. Faecal transplants.
5 Vrieze, A., *Gastroenterology* (2012); 143(4): 913–6. Transfer of intestinal microbiota from lean donors increases insulin sensitivity in individuals with metabolic syndrome.
6 Alang, N., OFID.2015.http://ofid.oxfordjournals.org/content/2/1/ofv004. full.pdf+html. Weight gain after Fecal Microbial Transplant; http://www. scientificamerican.com/article/fecal-transplants-may-up-risk-of-obesity-onset/
7 Charakida, M., *Lancet Diabetes Endocrinol* (Aug 2014); 2(8): 648–54. Lifelong patterns of BMI and cardiovascular phenotype in individuals aged 60–64 years in the 1946 British birth cohort study: an epidemiological study.
8 Everard, A., *Proc Natl Acad Sci* (28 May 2013); 110(22): 9066–71. Cross-talk between *Akkermansia muciniphila* and intestinal epithelium controls diet-induced obesity.
9 Zimmermann, A., *Microbial Cell* (2014); 1(5): 150–3. When less is more: hormesis against stress and disease.
10 Lang, J.M., *PeerJ* (9 Dec 2014); https://peerj.com/articles/659/. The microbes we eat: abundance and taxonomy of microbes consumed in a day's worth of meals for three diets.

Index